THE BREAKDOWN OF

PSYCHIATRY:

A CRITIQUE

DINU GANGURE MD, JD

THE BREAKDOWN OF PSYCHIATRY:
A CRITIQUE

Printed in the United States of America

Published by Blue Globe Press

ISBN 978-0-578-60683-5 (paperback)
ISBN 978-0-578-60684-2 (eBook)

SUITS

It's a bull market in psychiatry these days. A simple search on the Internet of the words "majority of psychotropic medications prescribed without a diagnosis" shows the fury in the eyes of the bull. Lots of psychiatric medications are being prescribed to many people without a justifying diagnosis. A ton of money is being paid in psychiatry.

Two gladiators fight in the healthcare field: the doctor and the healthcare administrator, also known as the suit. The doctor prescribes. The suit handles the money.

The relationship between the doctor and the suit can be cutthroat. On day number two of a hospital stay, the suit might approach the doctor of a suicidal patient and say: "Doc, the insurance denied coverage. How about discharging the patient?" In a long-term facility, the same suit might comment to a doctor of a patient ready for discharge after two years in there: "Doc, the insurance is still willing to pay. How about not discharging the patient just yet?"

Many suits don't have medical backgrounds. They may demand the doctors to do what makes more money regardless of sound clinical practice. Messages from suits range from a mild "Work faster, Doc, to keep the cost down," to a strong "Why in the world did you admit the patient without insurance?" Sometimes the suits may insist on using cheap medications, or reducing the time with the patient, or fragmenting the care. Under pressure from suits, the doctors walk a tense tightrope between quality of care and finances. At stake are the lives of the patients against a pile of money.

When suits and doctors struggle for power, they may go as far as engaging in shouting matches. In this type of battle, the party who cracks first often looks at quitting. Let's consider, for example, the suit that believes the psychiatrist is wasting time when no patient is in the room. Here, the suit undercuts an important part of being a psychiatrist: comparing data gathered from the patient with the standard of care—a critical analysis that needs to be justified through a solid written argument backed by the weight of the evidence. The tension between the competing interests of the patient, like survival and freedom, has to be resolved in a thoughtful, diligent way. To write an opinion in front of the patient without time to reflect on competing interests, without looking things up, without corroborating evidence, is hard to do. By comparison, judges resolve the tension among competing interests by taking time to reflect in their chambers, away from the courtroom.

To keep the prescribing pattern free of financial motivation, a layer of separation is needed between suits and doctors. An example of a layer of separation is the organizing of the doctors in a group, called the medical staff group. Unfortunately, this layer of separation is under pressure from suits, pushing to find themselves on both sides of the layer of separation.

On the doctors' side, the pressure is on by the suits hiring the chief doctor, as opposed to the doctors electing the chief doctor. On the suits' side, the pressure is on by the suits hiring a doctor in the role of a suit.

Both the chief doctor and the doctor in the role of a suit are in the position to blindside other doctors, because of the bond that usually forms between "peers." By speaking their professional language, the chief doctor and the doctor turned suit are in the position to covertly steer the prescribing pattern of other doctors in line with where the money flows.

The thinner the layer of protection between suits and doctors, the greater the chance of financial contamination of how medicine is practiced. No wonder sometimes a medical staff group is left with only one doctor, alone against a team of suits.

It's a Wall Street game for the suits. Some suits keep the speed of the mill running at a reasonable pace. Others act like pigs, as defined by Wall Street: pigs are risk-takers looking for big scores over a short time. Pigs don't perform their due diligence. They are impatient, greedy, and emotional about investments. They push doctors to take higher risks. The higher the risk, the more money the pigs make.

On Wall Street there is a saying: "Bulls make money, bears make money, but pigs just get slaughtered!" Well, when it comes to medicine, I haven't seen a pig slaughtered yet. At least not in psychiatry, my specialty. Pigs getting fatter, that's another story.

Clinical details in psychiatry are more subjective than in other disciplines. For example, a psychiatrist does not have an X-ray to prove a diagnosis beyond doubt. The absence of hard proof lets pigs run wild.

At the opposite spectrum of suits from the pigs are the sloths. They are so slow to act that patients can do just about whatever they want—if it would be up to the sloths. In the name of patient satisfaction, the sloths slow down almost anything. The sloths might even allow patients to doctor-shop inside the same healthcare organization—for the purpose of firing one doctor in order to see another, with the hope of receiving, for instance, a medication demanded against the advice of the first doctor.

Unfortunately, under the influence of the sloths, patients tend to regress. Sloths enable pathological behaviors, fostering poor choices by patients. All the while, sloths calmly watch the money pouring in for healthcare services. Sloths are cynical: they are aggressive in letting the money pour in, through slowing down the clinical process, which ends up requiring more healthcare services in order to compensate for the slowdown in reaching the clinical target.

While pigs rush, depriving patients of thorough care, sloths go to the other extreme, pampering patients to the point of deprivation of growth. Both pigs and sloths do it for money, indirectly pressuring doctors to be money-driven irrespective of the quality of care. For instance, when patients demand specific treatments that are marginally appropriate for the clinical condition but are moneymakers nonetheless, the pressure from pigs or sloths adds to the pressure from patients. Then, practicing good psychiatry becomes a challenge, as what is right in psychiatry is not a clear-cut matter.

From time to time, doctors give in to the pressure. As a result, patients end up with treatments remote from the standard of care. This way, pigs and sloths foster brain illness. They prevent recovery. They sacrifice health in exchange for money. All the while, they are insulated from malpractice liability, because the doctors are responsible for the medical act.

Under pigs or sloths, doctors do not operate in optimum conditions. Instead, optimum conditions require an administrator that is neither an impatient pig nor a pokey sloth. A good administrator is like a swan—a creature of grace, striving for a balance between what is good for the patient and what makes money. The well-being of the patient is on the line.

ADHD

When the patient is a child, the well-being of the child is supposed to be of utmost importance.

Child psychiatry allows about thirty minutes per follow-up appointment. Out of the thirty minutes, about five minutes are spent with the electronic medical record encouraged by the government, five minutes with the mandatory documentation for billing purposes, five minutes with the school record, five minutes with the caregiver, and five minutes trying to break the ice with the child by engaging through play. That leaves a paltry five minutes to actually treat the patient.

In child psychiatry it is well known, but rarely said, that many times the real patient is the parent, not the child. By saying "Doc, fix my child," the parent risks continuing the same behavior that made the child suffer to begin with.

In the world of mental health consumerism, the psychiatrist that does not give the child a diagnosis to the satisfaction of the parent may lose the parent as customer. Quickly diagnosing the child with ADHD, bipolar, or both becomes tempting. But the only "diagnosis" the child has may be the psychologically toxic environment caused by the parent.

Here is an example. A child presents with mood disturbance, determined later to be the result of witnessing arguments between the parents. These arguments are not disclosed to the psychiatrist at the first visit because the accompanying parent does not see their importance, or because the child is shy about discussing the arguing. In light of the presenting mood disturbance, the diagnosis of "bipolar" is considered. So, the child is put on a strong medication, like lithium.

The cognitive side effects of lithium could cause the child to lag behind in school. Consequently, a strict classroom teacher demands the pupil be evaluated for ADHD. The strict teacher fills a form for the psychiatrist, noting the difference between how the child is now and how the child "should be" according to the strict teacher. Subsequently, the psychiatrist gives weight to the form received from the teacher. It becomes a basis for putting the diagnosis of ADHD.

This way, the child can end up with both diagnoses of bipolar and ADHD. Yet, the only "diagnosis" of the child in this example is being stuck in the psychologically toxic environment caused by the parent. If anything, the chemical imbalance of the child is the byproduct, not the cause of what goes on at home.

Putting an accurate diagnosis is a difficult task. Hopefully there will come a day when a diagnosis is centered on the functioning of the child's brain in an adapted environment, not on the brain of the child alone. Far fewer children would then receive an ADHD diagnosis. After all, ADHD is already supposed to be a functional diagnosis, meaning dysfunction is required by the definition of the illness in the *Diagnostic and Statistical Manual of Psychiatry (DSM)*. Yet, dysfunction is relative to the environment, a matter that the definition is silent about: in an adapted environment the child is more likely to function than in an unadapted environment.

In general, when considering the adaptability of the environment for the brain to function in, the main "diagnosis" may be the refusal of the environment to adapt. This is similar to a patient in a wheelchair not finding a way to climb the steps of a church. Who has the main problem here, resulting in an inability to enter the church: the wheelchair-bound patient, or the faithful group that did not extend the courtesy of a wheelchair ramp?

✱

ADHD treatment seems to have become for psychiatrists what cosmetic surgery is for surgeons. When treating serious deformations, cosmetic surgery makes sense. But when addressing minor imperfections noticeable only by a patient seeking an imaginary ideal of perfection, cosmetic surgery slides into a kind of a joke. Likewise, treating ADHD with severe features is serious business. But minor imperfections that don't amount to a real diagnosis, by not interfering with functioning, do not need much intervention in a psychiatric sense. At most, these minor imperfections need an adapted environment.

Where to draw the line, to stop the epidemic of ADHD? Facing a patient's imaginary problem, the psychiatrist has a choice: to give in, or to keep it real. Critical thinking and commercial thinking are in conflict within the psychiatrist. The choice depends on the values of the psychiatrist.

In a bad economy the unemployment rate raises, inviting patients to regard minor imperfections as a diagnosis to blame unemployment on, and sometimes to claim disability on: "Doc, I have ADHD, do something"—discounting that in a good economy the conversation may not take place. In a good economy, patients with minor imperfections stand a better chance of employment, thus of functioning just fine, not even close to considering a diagnosis.

✹

A cause of over-diagnosing ADHD is missing what is in the differential diagnosis. For instance, under pressure from a patient demanding ADHD medication due to medication-seeking behavior, not considering the differential diagnosis of substance abuse may result in an over-diagnosis of ADHD. A harm of the over-diagnosis of ADHD is the use of powerful stimulants with addictive potential when not necessary. Another harm is the creation of "reverse disability"—centered not on the brain, but on outside factors, like the economy: a disabled economy makes the patient look disabled, despite the culprit being the economy, not the patient.

The amphetamine in ADHD drugs is a cousin of methamphetamine, an illegal drug with powerful negative effects on the brain. Instead of leading to a lasting correction of a chemical imbalance, the amphetamine can act like a quicksand, by giving the illusion of stability and comfort at first, only to lead to tolerance and addiction in the long run. Ultimately, the amphetamine can fry the brain cells.

Addictive stimulants enhance the performance of the brain, and can be compared with steroids enhancing the performance of the body. But a mere deficit in performance is not necessarily an illness. To justify the side effect risks of stimulants, like addiction and psychosis, the presence of an illness is warranted.

Let's look at one example when a mere performance deficit does not necessarily raise to an illness. We know that the job of an air traffic controller requires a distributive attention in order to coordinate the flight paths of various planes at the same time. For an air traffic controller with a natural tendency for distributive attention, the match between the brain and the need for distributive attention in the control tower results in a good functioning—the chemistry of the brain matches the environment well. In this case, because the dysfunction required by definition in ADHD is not met, ADHD is *not* diagnosed.

On the other hand, put the same air traffic controller with distributive attention in a rigid environment, like working on a monotonous assembly line, where a small variation from average is counterproductive, and dysfunction knocks at the door. In general, a rigid environment that fails to adapt raises the risk that an individual will be diagnosed with a psychiatric illness.

When not willing to adapt, society invites an over-diagnosing of ADHD through an arm's-length isolationist approach toward those who need an adapted environment to function well. However, ADHD should a valid diagnosis only when two conditions are met: 1) symptoms are present; and, 2) functioning fails in a genuinely adapted environment.

The second condition conforms to the idea of the diagnosis being centered on both brain and functioning, not only on the brain. But because functioning depends on the environment, how the brain functions in an adapted environment needs to be considered before a diagnosis of ADHD comes into play. Symptoms alone, with functioning present, amount to a noticeable oddity at the most, but no more.

In the absence of functioning, the question often skipped is: can the environment adapt to bring the functioning back? Once functioning is restored, the requirement of dysfunction in the ADHD definition is not met anymore. Thus, when the environment adaptation brings back the functioning of the patient, the diagnosis of ADHD goes away.

An unconcerned psychiatrist can ignore the need for the environment to adapt, and slam a diagnosis of ADHD by virtue of how the patient does in an unadapted environment. But a concerned psychiatrist considers how the environment adaptation may turn dysfunction into function, which leads to the elimination of the "diagnosis" of ADHD.

By comparison, a lactose intolerant person initially diagnosed with a stomach disorder has no more symptoms after avoiding lactose. The diagnosis of stomach disorder is downgraded to a mere warning: avoid lactose and you will be fine. Likewise, in the case of suspicion of ADHD, avoiding an environment that does not adapt may eliminate the diagnosis of ADHD, leaving only a warning in place: avoid the environment that does not adapt.

This being said, in a severe case, when dysfunction continues after the environment adapts, the ADHD diagnosis stands.

LENSES

This book is made of notes I wrote in a personal journal, on various days in the adventure of being a psychiatrist. Like different days of a week deal with different problems, chapters in the book deal with matters not always connected with each other. I've done what I could to smoothen the transitions.

As we begin the journey through this book, which I call a critique, let's look at the lenses by which to read it through. The lenses are like axioms in mathematics—not proven, but basic assumptions, on which the whole system is constructed. Here is a moment of mental gymnastics: there are mathematics built on the axiomatic assumption that $1 + 1 = 3$. From there, a world of fascination for mathematicians evolves, as long as one does not bother to ask how $1 + 1$ can be 3. In the new world, $1 + 1$ is indisputably 3. What may seem like an odd, out of ordinary assumption, later turns into the stepping-stone for the development of a surprisingly coherent mathematical universe. Likewise, in this critique I make two key assumptions, or axioms, indisputable by definition. They are the lenses, or the keys, by which to read the critique, which makes no sense without them.

The first assumption, or axiom, is that every person has a core need to belong. I hope scientific research will prove this axiom to be true. The critique does not set out to prove it. The critique merely assumes it.

The second assumption, or axiom, is that the monologue of a lonely brain has no more power than the brain itself, while two brains working together connected can be more powerful than the same two brains working simultaneously but disconnected. Because a single brain in monologue is no more powerful than the brain itself, I do not describe the lonely brain as having a "mind" of its own. On the other hand, because two brains working together connected can be more powerful than the same two brains working simultaneously but disconnected, I call the connection between the two brains the mind. Thus, it takes two brains to have a mind.

Science has shown that the psychiatric illness is in the brain, like a brain tumor, with the difference being in size: psychiatric illness tends to be microscopic, while brain tumor tends to be larger. Because science has established the psychiatric illness to be in the brain, the term "mental illness" is replaced with "brain illness" in this book. I realize the book breaks with tradition here, but if you are looking to keep the tradition, this is not the book for you. Here it takes two brains to have a mind, in contrast with one brain to have an illness. Therefore, the accurate description of the psychiatric illness is brain illness, not mental illness.

✳

Three powers are involved in a mind: P1 = the power of one brain; P2 = the power of the other brain; and P3 = P1 combined with P2. What the mind between the two brains can do depends on the first brain as well as on the second brain. It is a function of the two brains making up their unique mind. But P3 is not exactly the sum of P1 and P2, as their combination can add power. This critique assumes, in the second axiom, that a combination of two brains can be more than their simple sum.

A good example is the mind formed between a psychiatrist and a patient with autism. On surface, what two brains can do together appears limited by the lower-functioning brain. Conventional wisdom says that their mind—the connection between the two brains—can only go as far as the brain with lower function allows; for instance, that enacting a short scene from a theater play, as part of psychiatric rehabilitation, cannot be accomplished when one brain cannot do it.

This critique sets to show how, despite the limitations of the brain with the lower power, the brain with the higher power holds the key to empowering the mind. Through an empowered mind, the less powerful brain, even when left behind previously, can find a way back to a higher power.

✳

I happen to regard the bundle of our relationships to be central to who we are. On the contrary, psychiatry over the years has pushed to dampen the importance of relationships, by looking at the human as an independent impersonal machine, self-sustaining and independent of others. The disaccord is not of logical nature, but emotional: psychiatry claims a rational approach, while I am comfortable with an emotional belief. In my view, psychiatry needs rationalization to shy away from the less profitable perspective that relationships are at the core of who the patient is.

When relations matter less, what counts more is the number of neurotransmitters in the brain. The quest for healing becomes a search for chemicals, not a search for chemistry. Statistics, numbers, pills, electric shocks, magnetic currents take the stage. On the same stage, not much room is left for soul-searching.

Drawing for a moment a comparison with a game of chess, psychiatry does not focus much on the strategy of the game. Instead, psychiatry focuses on the pieces on the chess board: what they are made of, how strong they are, how to glue them back when broken, and what they can do for a basic move—for instance, that a horse moves in an L shape.

Somehow, the field of psychiatry shifted the attention of the society from the strategy of the game to the sturdiness of the pieces. Attempting to respect all game styles, psychiatry does not promote one strategy over another. Curiously, money too is style-blind, not promoting one strategy over another. Merely being in the game gives the appearance of being on track. But patients need more than being in the game. Patients need winning.

✳

A homeless person sharing the sky with another homeless person is not alone; when engaged in a friendly conversation, they belong to a narrative, a story together. Then, the government steps in and gives them housing, miles apart from each other. It seems a step forward— walls, light, and heat are now in place. But with homelessness now gone, the new enemy is loneliness.

The previous sharing of the sky on the street was psychologically more powerful than the loneliness in the new housing. The mind on the street is now replaced by the mindless state of being alone. With nobody to say "Welcome home," the house is empty on the inside.

When the rules of subsidy for government-assisted housing forbid living together, no wonder some people prefer returning to a shared homelessness than living alone.

＊

Of the people who settle in the new housing, many struggle to belong to a shared narrative again. This can open the door to depression. The word "depression" has two sides. On one side is the lay use of the word "depression," often describing merely a sad feeling. On the other side is the clinical use of the word "depression," going to the depth of symptoms beyond a sad feeling, and accompanied by loss of functioning. Symptoms beyond a sad feeling can be too little or too much sleep, too little or too much appetite, low self-esteem, hopelessness, helplessness.

When everybody wants to be an expert in psychiatry, the job of the psychiatrist can be difficult. "You are depressed," says the psychiatrist. "Oh, I know what you mean!" How to tell the difference between the lay use of the word "depression" and the clinical use is not always easy. An individual in denial of having clinical depression might try to explain away each symptom, separately from the rest of the symptoms. But taken together, the combined symptoms act as the pieces of a puzzle: they have the force to unfold unequivocally the image of a clinical depression. What seemed to be a mild gust of wind turns into a tornado—a clinical depression that can tear apart achievements, relationships, and even lives.

Just as the word "depression" has two meanings—the lay meaning and the clinical meaning—so does the word "recovery" have two meanings: recovery of the brain and recovery of the mind. Merely because the brain is chemically rebalanced (recovery of the brain) does not mean the mind comes back to life (recovery of the mind).

A depression that does not kill the brain can very well kill the mind. Waking up from depression through chemical treatment (recovery of the brain) can still leave the patient in a mindless state (no recovery of the mind). The recovery of a mind is a rehabilitative process that goes far beyond a chemical rebalancing of the brain.

Like the fusion of two atoms discharges energy, a mind fills the air with the power of connection. A mere example of connection is the power of love. Within a mind in love, the brain comes to a new life. What makes the brain work then is not only the biological power of each of the two brains; the mind in love becomes a distinct source of power for the brain.

In general when a brain is about to fail, the mind offers protection. A failed mind deprives the brain about to fail of the protection that the mind would otherwise offer. With a failed brain and a failed mind, death comes near.

SUICIDE

The idea of suicide is ubiquitous. It exists in the abstract—on a piece of paper, on the Internet, in history. While the idea of suicide can be present in anyone's awareness, the vast majority of people have strong enough brains to reject it. On the other hand, a vulnerable brain due to illness can accept the idea of suicide, instead of rejecting it. Note that suicide involves the *idea* of suicide. Without the idea of suicide, death is an accident at the most, but not a suicide.

A debate revolves around whether a suicidal person is morally guilty. The modern view is that suicide is not the fault of the person who commits it. But whose fault is it then?

Medicine does not usually assign fault. Instead, medicine hunts for a culprit. The evil idea of suicide, which finds a home in the ill brain, is exactly that—plain evil (the culprit). On the other hand, without assigning fault, the vulnerability of the brain due to illness (the facilitator, not the culprit) makes the brain incapable of rejecting the culprit (the idea of suicide).

Having a vulnerable brain due to illness, however, is not a mandatory condition for suicide. Even brains of great strength have killed themselves. This keeps alive the moral dilemma of fault in suicide: is there fault in suicide in the absence of brain illness?

Suicidal behavior is the end manifestation of different factors that overlap in various degrees.

On one extreme are the suicides of people without control over what they're doing. For instance, consider a very depressed person who struggled for years to reject the idea of suicide. As the brain succumbs to depression, the idea of suicide works its way in. The severe depression makes the brain powerless to reject the idea of suicide. Instead, the idea of suicide catches on, followed by self-inflicted death.

At the other extreme are people in full control of what they're doing. Take, for instance, the terrorists on September 11, 2001. Their suicides weren't driven by brain illness, but by a voluntary submission to a doctrine of destruction.

The vast majority of suicidal people struggle to control behavior. They are in the gray area of battling against suicide, in the tension between the good idea of survival and the bad idea of self-inflicted death, with the brain in a fight mode for survival.

Jumping to a conclusion of what led to an apparent suicide may not be accurate, being reductionist instead. This is like jumping to a conclusion that a driver is at fault for hitting a tree. But did the car have faulty brakes, maybe? Instead of rushing to a conclusion, key to accurately understand an apparent suicide is analyzing on two separate dimensions what happened:

> Dimension 1: The predominant idea—ranging from the suicidal idea being predominant to the survival idea being predominant;

Dimension 2: The power of the brain—ranging from the brain having low power due to illness to the brain having high power due to health.

Here is a simplified table illustrating the likely outcome when the suicidal idea is predominant:

Suicidal Idea Predominant	Brain Power	Likely Outcome
strongly predominant	low (ill)	suicide
strongly predominant	high (healthy)	toss-up
weakly predominant	high (healthy)	survival
weakly predominant	low (ill)	toss-up

For comparison, here is a simplified table illustrating the likely outcome when the survival idea is predominant:

Survival Idea Predominant	Brain Power	Likely Outcome
strongly predominant	low (ill)	survival
strongly predominant	high (healthy)	survival
weakly predominant	high (healthy)	survival
weakly predominant	low (ill)	survival

The two tables above show the critical role played by the ideas flowing through the brain in whether the individual lives or dies. As seen in the second table, a predominant survival idea results in a likely survival outcome across board. As seen in the first table however, a predominant suicidal idea results in a variable outcome—from suicide, to toss-up, to survival—under the influence of whether the brain is healthy or ill.

Assessing the safety to self involves determining which idea is predominant—survival or suicidal—as well as determining to what extent it is predominant—strongly or weakly.

Because brain illness tends to fracture what the patient says from what goes on in the patient's head, the determination of which idea is predominant, and to what extent it is, takes quite a bit of detective skill from the psychiatrist, far beyond asking the patient: are you suicidal?

✳

A suicidal idea can be predominant in two ways: by illness, and by choice.

In a "medical" suicide, the brain is unable to reject the idea of suicide due to brain illness—a non-autonomous, automatic process, without self-control. An example here is a brain with psychosis leading to completed suicide.

By contrast with the "medical" suicide, I will call the instance where a healthy brain chooses to commit suicide a "forensic" suicide—an autonomous process, within self-control. An example here is of the prominent Nazi members that made the choice to self-inflict death instead of being held accountable at the end of World War II.

Thus, on one side of the spectrum is the medical issue of the inability to reject the idea of suicide (an ill brain). On the opposite side of the spectrum is the forensic issue of the choice of the idea of suicide (a healthy brain).

Just as apples and oranges are different, a brain illness and a choice of idea are different too. Likewise, a medical suicide is different than a forensic suicide. In a medical suicide, the brain has an illness, a process defect without choice. In a forensic suicide, the brain makes a substantive choice, of the idea of suicide.

While a forensic suicide does not require a brain illness, in a medical suicide the brain illness is a facilitator of death—creating the opportunity for the idea of suicide to stick to the brain, instead of being rejected by the brain. Note that the idea of suicide is the cause of death in both forensic and medical suicide.

Assessing what led to the suicide includes a forensic study of the ideas embraced by the individual and a psychiatric analysis of the case. While the individual is responsible for a forensic suicide, is the patient responsible for a medical suicide? When a construction worker chooses to not wear a hard hat (a self-induced matter, the equivalent of stopping the meds), is the construction worker not responsible for the injury that follows? (is the patient who chooses to go off medications not responsible for the medical suicide that follows?)

＊

In a case at risk for suicide, changing the outcome from death to survival requires tackling the imbalance between risk factors and protective factors: reducing risk factors, enhancing protective factors, turning risk factors into protective factors. The factors involved, whether risk or protective factors, can be biological (chemical), psychological (internal process of thinking and feeling within the brain), social (relations with people), and doctrinal (relations with ideas).

In today's society, psychiatry is expected to take care of the biological, psychological, and social factors. What is missing from the picture is a doctrinal manager, to handle people's relations with ideas. Psychiatry does not promote a specific doctrine of ideas beyond freedom of ideas. It's like a gardener that does not cultivate a specific tree, but allows almost every weed under the sun to grow, despite the likelihood of ending up with a poisonous tree later.

In a world where ideas can be more powerful than brains, the doctrinal manager becomes a critical component of fighting against suicide. A survival doctrine gives power to a brain on the brink of suicide to stay alive, just like a column of raising air adds power to an airplane to fly without engine. When everything else fails, the column of raising air can save lives.

✳

Every day, lawmakers ask the psychiatrist: Is the patient suicidal? The lawmakers like black-and-white answers. But patients can be both suicidal and not suicidal at the same time: some patients are ambivalent about suicide, contemplating suicide but not ready to go through with it.

How to help patients that are ambivalent about suicide stay alive? Acknowledging that there is such a thing as ambivalence about suicide is the first step. Then, understanding what's on both sides of the ambivalence about suicide is important.

Despite risk factors of suicide, enough protective factors against suicide can maintain alive the individual ambivalent about suicide—in a rather fragile equilibrium nonetheless. Sometimes turning a single factor from protective into a risk factor can be enough to imbalance the individual toward completing suicide. Take for example the protective factor against suicide of the strength of the patient's marital relationship. When divorce comes up, the marital relationship can easily switch from being a protective factor to a risk factor. This can be enough to move the patient from suicide contemplation to suicide completion.

For patients on the suicide fence, who contemplate suicide but do not yet act on it, time is of essence. An intervention may come too late, as the freedom to be on the fence is a false friend. When on the suicide fence, despite no overt signs of impending completion of suicide, the danger is already there. Remember the expression "It's all in your head"? Well, where else could it be? On the suicide fence, the idea of suicide is already at play, putting the patient in clear and present danger. From there, the time left to the completion of suicide can be awfully short.

The psychiatrist is called upon to facilitate the patient's rejection of the idea of suicide. In that, the psychiatrist delivering ideas that strengthen the patient together with delivering psychiatric care for the brain is a stronger approach than delivering only psychiatric care for the brain.

*

What ideas has the psychiatrist within reach to deliver, in order to strengthen the patient? The psychiatrist starts as a scientist, restricted to fact-finding and logically driven conclusions. On the other hand, the psychiatrist becomes less of a scientist, and more of an artist, when creatively employing ideas to problem solve with the patient. Because ideas can take a myriad of forms, the science of psychiatry turns into the art of psychiatry.

To transition from being a scientist to an artist, the psychiatrist enters in conflict with the constraints of science. But contrary to the logic of inference and deductions, the psychiatrist has the power to spread the wings of ideas—a taboo of science. Like an artist, the psychiatrist brings to the table ideas unsupported by science. There is a small, but essential nuance here: ideas unsupported by science are not necessarily contradicted by science.

*

I heard someone comparing the psychiatrist with a luthier. Like the luthier repairs a musical instrument, the psychiatrist repairs a brain. But no musical instrument is complete by itself, without music. So too, a brain is not complete by itself, without function.

This book stands for the proposition that the psychiatrist is called upon to be more than a luthier. The psychiatrist is called upon to be a musician.

✺

When the psychiatrist is a musician, an apparent conflict comes up with Uncle Sam, who many times pays the bill. What music is Uncle Sam exactly paying for? An answer on surface is that Uncle Sam is paying for the time of the psychiatrist. But will Uncle Sam be happy with the music chosen by the psychiatrist?

Let's take the example of Uncle Sam dealing with a Christian psychiatrist, when Uncle Sam pays the bill. How will Uncle Sam reconcile the non-Christian views of the state with the Christian views of the psychiatrist?

The Christian psychiatrist can put patients on advanced notice of the Christian views, so the patients can make an informed choice on whether to show up for the first appointment. Believing that what belongs to God is due to God, the Christian psychiatrist does not see oneself as the know-it-all guru who steals the show. Instead, the Christian psychiatrist acknowledges the limits of psychiatry. Will Uncle Sam acknowledge its own limits?

Uncle Sam paying for a psychiatrist's time is not the same with Uncle Sam having the right to limit the psychiatrist' ideas when facing the collapse of the patient's inner world. Ideas of the psychiatrist that bring strength against the structural shortcomings of the patient's brain can make the difference between life and death, knowing that the abyss of negativity secondary to the dysfunction in illness invites to despair, hopelessness, and fear.

To compensate against the abyss of negativity, a source of positivity is necessary. By the doctor-patient relationship, the psychiatrist is in the position to guide the patient to a source of positivity through ideas. Uncle Sam, on the other hand, is called upon to facilitate healing, not to jeopardize it. When the patient is the musical instrument and the psychiatrist is the musician, one would hope that Uncle Sam will refrain from the ambition of becoming a composer. Music does not need Uncle Sam in order to exist.

✳

Ideas take root and change behaviors. Ideas grow within the person, and lead to the transformation of fantasy into reality. Violent video games, as good as they might be for discharging aggressive impulses in a virtual environment, tear at the same time at the barrier between imaginary and reality. In people with a preexistent weak barrier, imaginary and reality begin to blend, the barrier gets weaker over time, and the imaginary may end up taking over behavior in reality.

While nobody can stop a person from having ideas but oneself, the psychiatrist is in the privileged position to challenge ideas—because the psychiatrist is a virtual tool to self-explore through mirroring, to a limited extent, what is in the head of the patient.

Good ideas coexist with bad ideas in the head of the patient, who is called upon to choose what ideas are worth keeping and what ideas are worth discarding. When the patient is not able to sort this out alone, the psychiatrist can challenge the coexistence of good and bad ideas, by distorting how the psychiatrist mirrors what is in the patient's head.

A psychiatrist mirroring accurately what is in the patient's head, without distorting, can be detrimental to the patient, by fostering simultaneously good ideas and bad ideas. This can be the case of the silent psychiatrist. To show the patient a direction, the psychiatrist needs to talk.

SHOWTIME

With the patient having a choice of psychiatrists, which psychiatrist is worth returning to? What ingredients make for a good psychiatrist?

Well, an ingredient is an attractive performance by the psychiatrist. Like in pay-per-view on TV, a not-so-attractive performance can kill the cash flow. To keep the cash flowing, putting the patient on the hook through an attractive performance is desirable. How the psychiatrist attracts the patient can vary, but usually the psychiatrist makes the effort to find common ground with the patient.

Let's use the example of a sociopath who comes in for treatment of anxiety. When the psychiatrist cures the anxiety, the sociopath is happy with the psychiatrist, even though the sociopathic qualities may *increase* as a result of the cured anxiety. When, instead, the psychiatrist jumps to tackle the sociopathic core, the patient might reject the psychiatrist, because the patient wanting no more than a change in anxiety.

To a limited degree, the psychiatrist needs to avoid challenging the inner indiscipline of the patient. Instead, the psychiatrist needs to invite the patient to find oneself a little bit in the distorted mirror represented by the psychiatrist. This gives the patient a sense of familiarity, which brings along a sense of security. In turn, the security leads to trust, which compensates against the fear of crossing through treatment the psychological limits of illness.

A psychiatrist distorting the mirror beyond the patient's self-recognition threatens the treatment progress—the patient does not find oneself within the too distorted mirror represented by the psychiatrist. This wakes up the patient from inner exploration. Now the psychiatrist prevents the patient from looking inside. Consequently, the patient may look outside, at the psychiatrist. Then, the work of the patient moves from being done internally, within the patient, to externally, between the patient and the psychiatrist. This may be appealing from the human development point of view, but can be a slowdown from the psychiatric treatment point of view.

A psychiatrist distorting the mirror too much facilitates an increase in treatment resistance. This paves the way for the patient to disengage from treatment. Thus, how far the distorting of the mirror in treatment can go is limited.

＊

Sometimes instead of mirroring the patient, the psychiatrist must put a stop to where the patient is going. "Give me some Xanax, Doc!" the patient says, as though talking to a bartender. In turn, the bartender might ask: "On the rocks?" Fortunately, the psychiatrist is not a bartender.

There are similarities between bartenders and psychiatrists. Both talk nicely to clients. Both make clients feel appreciated for showing up. Both try to boost the self-esteem of clients. And both mix ingredients to help clients feel better.

And yet, the psychiatrist must do what keeps the patient healthy, not what keeps the patient happy. Psychiatry in general is not about finding happiness, but about finding health. A tension is present between health and happiness. Early in training, a psychiatrist learns to say no, while communicating at the same time validation and support to the patient. The psychiatrist is not supposed to be a drug dealer with license.

Achieving client satisfaction can be hard when following the letter of medicine. This can be even harder in psychiatry, when the instrument to measure patient satisfaction is impaired—the patient's brain.

＊

Health and consumption of healthcare are two different things. The goal of achieving health is at risk of turning into the goal of consuming healthcare when professional organizations regard psychiatric patients as consumers. The difference is subtle, like between walking home and just walking.

Calling psychiatric patients consumers diverts the attention from the illness as driver of treatment to what the patients want to consume.

But patients are not free to consume what they want, as the illness, outside control of patients, determines what treatment is needed.

Without control over illness, the psychiatric patients look for control over at least something. Control over treatment becomes tempting. Treatment demanded by psychiatric patients but marginally appropriate for the clinical condition can be a dangerous pitch by professional organizations, as the treatment can slide remotely from the standard of care. Nonetheless, it can still happen, to keep the money flowing, making bean counters happy.

✸

Healthcare operates in a business model, where the need for revenue invites bean counters between patients and doctors. Bean counters like simplicity. They like things that can be counted.

How can psychiatry be packaged in a vehicle amenable to counting? Encapsulating life dimensions into a diagnosis, and encapsulating a patient's needs into a treatment (be it a physical capsule in the form of a pill, or a capsule of time in the form of a therapeutic session) create simplicity, which can be counted. Yet, a patient's life may be one "strike" to the end, with survival at stake.

The psychiatrist is like a baseball coach in the bullpen, with the science of baseball in hand. When time is running out, the psychiatrist better become creative, outside the capsules driven by the economy of psychiatry.

✸

While a good psychiatric treatment can turn a patient from a poor batter to a home-run hitter, no friends or family may be left in the audience to applaud the home run, having already been disappointed by the patient. Brain illness can poison the relations with family and friends. Without an audience, the batter is surrounded by the sadness of silence, threatening the efficiency of the treatment despite the home run. The alienation of the minds formed by the ill brain can be a separate risk factor in brain illness than the brain itself.

As the surrounding minds can go from none to many, and as the alienation of the surrounding minds can happen in various degrees, assessing the ill brain in light of the surrounding minds is not a one-size-fits-all operation. Instead, it becomes an out of ordinary operation, where patterns don't survive easily, and where fighting barriers, seeing where nobody saw before, and going above and beyond can be necessary to help the patient.

Attempting to encapsulate life dimensions and the patient's needs into countable items, to turn psychiatry into a financial vehicle, steals from the healing potential of psychiatry. Life dimensions are more than a diagnosis. What the patient needs is more than a capsule (be it a physical capsule in the form of a pill, or a capsule of time in the form of a therapeutic session).

No wonder the psychiatrists fight back the bean counters, by offering a large variety of services, pushing away the simplicity liked by the bean counters.

✳

A long time ago, psychoanalysis ruled the profession of psychiatry. Nowadays, the age of the attendees at the meetings of psychoanalytic associations is skewed toward retirees, with few young doctors in attendance. It is a sign of how far the priorities have changed in psychiatry since psychoanalysis was at the top.

While psychoanalysis required patients to lie down on the couch multiple times a week, more recently brief psychotherapies propose a treatment that involves sitting on a chair once a week.

But taking center stage, psychopharmacology emerges as today's hot runway model. The buzz is in the meds. A majority of psychiatrists do not expect a conversation with their patients multiple times a week, not even once a week. Instead, a once a month medication check may be considered enough.

Then there's the electric shock therapy, ready to put electrical wires around the heads of the patients. It comes down to a push of a button, for the current to flow. What talk? What introspection? What exploration of feelings?

All the psychiatric services make for a crowded place. One thing they have in common is making money off the patient. There's no business like the psych business.

BET

When selecting what treatment to offer, the psychiatrist has scientific methods to rely on. But when it comes to behavior, the patient, not the psychiatrist, is at the driver's wheel. Moreover, the patient's behavior is subject to illness, a non-autonomous process, not within self-control. The overlap between the patient being at the driver's wheel and the patient's behavior being subject to illness puts the psychiatrist in a risky business. It's a matter of luck.

The operations of a mental health organization remind me of a casino.

In managing the luck, the pharmacy of the mental health organization is like a card dealer in a casino, delivering the "cards" (the meds). Mistakes by the pharmacy are like handing out the cards in wrong order, jeopardizing the likelihood to win.

The executives in the mental health organization are like the floor supervisors in the casino. They want gamblers at each table. They may throw in a lunch, a movie pass, or even a day trip, to keep the gamblers gambling.

Whether gamblers win or lose, at the end of the day the house collects. The mental health organization is insulated from serious financial dents by the liability of gambling being attached to the psychiatrist, and by having deep pockets that can absorb an occasional loss.

While there are similarities between a mental health organization and a casino, there is at least one difference. The gambler knows that the gambling machine is supposed to work well. But the psychiatrist gambles on a machine that usually does not work well, by definition: the brain of the patient, where dysfunction is usually required, by definition.

✳

An elevated way to say that psychiatry is a matter of luck is that psychiatry is a matter of risk management within the confines of statistical significance.

This reminds me of a poster advertisement for a theater play based on Tolstoy's novel *War and Peace*. The poster shows a long table, with no food at the ends, and lots of food in the middle. Rich folks sit in the middle, with a candle lit nearby, eating ferociously, and glancing in surprise at the ends of the table. Soldiers in gear of war fight at the ends of the table, where no food is present, no candle, and not even tablecloth. The soldiers would rather eat something, but with no food available to them, fighting is what they are left to do.

One interpretation by the audience was that the poster illustrates the theme of the ignorance of the rich toward the poor. Drawing a comparison, in psychiatry the people at the middle of the table decide the funding, and consequently influence what happens in psychiatric care through the power of the purse. From the middle of the table to the ends, the decision-making process trickles down in a controlled fashion, through "psychiatric care."

When patients ask from the sides of the table "Can I get some food?", the flow of the money decides what the patients get. When the flow of the money pays for therapy, patients get therapy. The food requested by the patients is a metaphor for the riches at the middle of the table: electricity, heat, a phone, a washer, a coat, a bus ticket. This is how the therapy session might go. The patient says: "I am hungry." The response of the therapist: "Tell me how you feel. By exploring feelings, you can better manage what's bothering you."

Despite the illusion of care through the smoke screen of therapy, "hungry" patients need more. Food, even when being a metaphor for the riches, is more than food for thought by way of the therapy.

✳

If unable to meet the patients' needs, mental health organizations get smart. Let's consider, for instance, the use of the words "mental health" in their title. This is done despite the illness sitting in the brain, and thus "brain health" being more accurate in the title of the organizations than "mental health."

The mental health organizations are businesses that move money, even when designated as nonprofit. For some of them, revenue is tied mainly to putting forth a diagnosis, but not so much to pursuing an improvement in the ill brain. Then, after putting forth a diagnosis, these organizations are left to merely adapt, by client apple-polishing, to the ill brain—adapting being financially more efficient than pursuing an improvement of the ill brain.

In general, the ability of a healthy brain to adapt to the dysfunction of an ill brain is a good thing, so the ill brain can function within the mind—adapting usually works in favor of the ill brain. For instance, by adapting to a child with autism, the parent allows the child to function through the parent-child mind, which works in favor of the ill brain of the child. But for a mental health organization to merely adapt to the ill brain of a patient, instead of pursuing an improvement in the ill brain, that is a different story.

Scratching the patient's back can mask brain illness, by offering the appearance of health. After all, the patient is happy, so mental health is in place. The problem is that mental health is not brain health: when the patient steps outside of the mind formed with that who scratches the back of the patient, brain illness stands the chance to go unmasked.

MEDICATION

To pursue brain health, medications might come into play. While solid professional boundaries are in place between the psychiatrist and the patient due to valid ethical concerns, there is one place where the boundary is loose: prescribing medications. The psychiatrist has an extraordinary power to pile up medications on top of medications, for the same patient.

Scientific research shows that the lifespan of patients on multiple psychiatric medications tends to be shorter. Despite the shorter lifespan, the well-publicized trend is now to increase drug use for treating brain illness. Reasons for this trend include:

1. Drugs make brain illness look simple: a piece is missing, put the piece back (the chemical missing), problem solved;
2. Drugs can be counted, and paid for accordingly;
3. A drug can be replicated in the same format for everyone taking it, allowing a non-individualized intervention, standardized across board;
4. The brain rewires itself around psychiatric meds, making it difficult, if not impossible, to come off multiple simultaneous psychiatric meds. Attempting to do so may result in a recurrence of symptoms;
5. A psychiatrist who tries to prescribe one medication at a time may be regarded as weak by patients, who may run to a different psychiatrist in order to get the instant gratification of three or four meds at a time;
6. The isolationist agenda of the society tends to reduce patients to an illness: it must be the brain illness that needs to be addressed, not the person behind the illness;
7. The monetary value of conducting therapy is seriously diluted for psychiatrists when professionals who are not psychiatrists, such as social workers and psychologists, conduct therapy at much lower prices than the psychiatrists once did. It prevents psychiatrists to earn a sizeable return against the steep investment of going to medical school. Subsequently, psychiatrists find themselves drifting to a corner where money is—prescribing medications;
8. The notice sent by the government to stop the number of simultaneous medications being prescribed to a patient is practically toothless, being merely advisory;
9. As opposed to chaining a patient to the psychiatric bed, chemically restraining the patient by prescribing multiple meds is not apparent to the naked eye, and thus not startling the advocates of freedom. Calling chemical restraining by the name of "treatment" makes it even less visible.

Who needs personal growth when there is a "feel-good" pill out there? The trap is tempting. After all, many patients do feel good after taking the pill. The danger? The patients are stuck on relying on the pill to feel good. When encountering adversities in life, the patients face an increase in dose. This comes awfully close to the patients using street drugs to "feel good" by chemical sensation, bypassing the growth necessary to achieve accomplishments to feel good about.

When medications are used for the direct purpose of making a patient "feel good," they bypass the functioning of the brain, shortcutting the brain's opportunity to engage with the environment in order to feel good. On the other hand, restricting medications to merely freeing up the functioning of the brain from the illness favors the brain to engage with the environment in order to feel good.

The role of medications is not to take over the function of the brain. Medications are supposed to enhance the functioning of the brain, not to replace it. When medications bypass the functioning of the brain, they ultimately perpetuate dysfunction, becoming part of the illness, not treatment.

Compare this with illicit drugs, a restrictive universe that bypasses the functioning of the brain by directly making the brain feel good chemically, through hijacking the pleasure circuit of the brain from routine, ordinary life experiences.

✸

Some patients wonder whether they should turn to God instead of medications. The answer comes from God's wish for the human beings to function. Thus, when medications are used to enhance the patients' engagement with the surrounding environment in order to function, God and medications are compatible. When, however, medications are used to bypass functioning, directly making patients feel good chemically, then God and medications are not compatible.

Whether medications and God are compatible depends therefore on the purpose of medication use: to enhance functioning versus to bypass functioning.

✸

Once in a while a patient comes across off medication and behaving like a zombie. After taking medication, the same patient says: "the medication makes me feel like a zombie." So, what is worse: behaving like a zombie, or feeling like a zombie?

In finding the right medication balance, the psychiatrist escapes the constraints of an exact science, as the balance is not the same for everybody. But one thing is for sure: the psychiatrist is not supposed to be a prescribing zombie.

✸

When benefits outweigh risks, medications are a good thing. Like open-heart surgery, medications are meant to save lives. But also, like open-heart surgery, medications carry serious risks. The inserts packages of medications talk about risks, like kidney failure, aggression, blood clots, stroke, liver failure, and more.

In the decision to medicate, the psychiatrist applies medical statistics to an individual patient. How to apply statistics to an individual patient centers on the psychiatrist's belief of "what is good" for that particular patient, which in turn is shaped by the psychiatrist's own values. For instance, a psychiatrist might believe that a bit of ongoing anxiety gives an edge to the patient that enhances functioning, while another psychiatrist might consider that a bit of ongoing anxiety is a problem deserving treatment.

＊

This brings us to a poster advertisement from Big Pharma for a psychiatric medication, showing on the left side a patient alone in the middle of a storm, and on the right side the same patient with a beautiful girlfriend, on a sunny day, in the park. Seemingly, the poster shows that, upon taking the medication, the patient moves from the loneliness of the storm to the joy of the relationship.

How the girlfriend popped out of nowhere is not clear. The advertisement can be read in reverse: the girlfriend was there first, on a sunny day, in the patient's imaginary, while the patient was off meds. After the patient took meds, the weather got bad: the patient woke up to the reality of the loneliness.

"Take this pill and you will be fine" is one of the worst lines that a psychiatrist can say. Pills are bridges, not destinations. Pills are important, because crossing from illness to health may not be possible without them. But, just because pills can be necessary is not the same with pills being sufficient.

Often enough, patients wake up from brain illness to realize the dearth of the surrounding reality. While the strength of the patient induced by medication may be useful in the task of building a more favorable reality, the recovery process does not end by taking the pill. It merely begins. "Take this pill and you can become stronger" is a better line than "take this pill and you will be fine."

Besides medications when needed, what is good care for a patient with brain illness?

＊

Well, for one, we know it takes a village to raise a child. Does it take a village to care, alongside the psychiatrist, for a patient with brain illness? Or, can the village care for the patient with brain illness by simply delegating it to a psychiatrist?

Once in a while an extreme act by a patient, like hurting others, hits the news. Then, community members ask dumbfounded whether more money should be put in psychiatry. Delegating to a psychiatrist becomes tempting. The community contemplates paying a way out of the problem.

However, by delegating to the psychiatrist, the community deprives the patient of the many minds that can be formed by the patient in the community. Before actually hurting of others, a patient at risk of hurting others can benefit from fully developed relationships with family and friends, in order to share values through connections, and ultimately to avoid hurting others. The patient needs a story to belong to, with characters, dynamics and a net of relations, where a node matters to other nodes and it can invoke the help of the entire net if needed.

To keep the patient safe, minds can be instrumental. Consequently, the community cannot afford to deprive the patient at risk of an extreme act of the minds in the community, by delegating to the psychiatrist. Instead, the community, or the village if you wish, is called upon to become part of the care.

Within the village, the basic unit of social organization is usually the family.

FAMILY

How to deal with brain illness within a family?

Members of the family negotiate peace between them by reaching a balance among family interests. Usually, family members prefer to rely on predictability. When a family member has brain illness, however, not only predictability, but surprises occur too. The surprises from the family member with brain illness are not always pleasant to the other family members, and can result in strains within the family, or even ruptures. Then, the family member with brain illness may ultimately fall out of the family narrative.

The job of the psychiatrist is to reconnect the family member with brain illness to the family narrative, to the story of that family. Of course, this is easier said than done. Reconnecting a patient requires focusing on the family narrative, not only on a chemical imbalance within the patient. For the reconnecting to the family narrative to happen, however, more than the patient and the psychiatrist have to agree: the family has to agree, too.

While the psychiatrist has methods, techniques, insight, knowledge, and an emotional distance from the patient, the family members can have a hard time not to take personally the surprises from the patient. Temporarily redefining the relationship with the patient as impersonal becomes important in the road to recovery for the family. This way, the family members can regard the patient's surprising behavior to be driven by a non-autonomous, automatic process of illness, outside self-control, and not by choice.

This is a hard "pill" to take in for the family. "What do you mean, Doc? So, you're saying the patient did not throw an object into the wall, but the illness did?" For family, it's always been personal—that's what makes it a family. Redefining the relation with the patient as impersonal strikes at the very core of the family. Overcoming the patient's surprising behavior can be exceedingly hard for the family.

Resilience is necessary to restore the family equilibrium. Brain illness is a threat to the minds formed between the patient and other family members. The patient is caught in the tension between the need to belong to a family narrative and the propensity of the ill brain to sabotage the belonging. Holding on to the relationship with the patient requires family members to infuse in the relationship emotional capital in advance, without an expectation to always get a return.

＊

Members of a family can summon each other in order to overcome obstacles together. Note that, when a patient is disconnected from family, the need of the patient to belong to the family remains in place. This is valid even when no family is left.

Where do lonely patients go? Sometimes they may go to a place like Fountain House, a community-based social club targeting the recovery of people with brain illness. Fountain House is spread today under different names in hundreds of locations all over the world.

A sense of belonging is present at Fountain House, a sense of journeying together—patients and staff. Similar to how folks in ancient times gathered around the warmth of a fire ring to share a bond, patients and staff gather together at Fountain House to fight the darkness of loneliness. The conversation shared at Fountain House gives hope. The sense of community becomes meaningful to patients for satisfying the need to belong to a story.

In a rational reality, Fountain House is a place where patients and staff get together for activities such as vocational rehabilitation, cooking, exercising, working on computers, socializing, and negotiating how to get along with each other. In an emotional reality, Fountain House may be the only family known to patients. When belonging nowhere else, Fountain House is the patients' narrative.

At Fountain House, searching for normalcy takes place together, patients and staff, under the same name: members. As a first step to normalcy, nobody wears badges at Fountain House.

NORMALCY

It is easy for a psychiatrist to describe what illness is. *The Diagnostic and Statistical Manual of Mental Disorders (DSM)* is a psychiatric manual containing plenty of information about what illness is. However, psychiatry does not have a *Manual of Normal*. What is, then, the normal?

Seemingly, the normal is the absence of a psychiatric diagnosis. But the absence of evil is not the presence of good. Thus, when the normal is squeezed to fit within the absence of a psychiatric diagnosis, a piece is missing.

What piece is missing? If you ask the government, the consideration between the government and the patient is financial—like one dollar, one pill. So, maybe money is missing? Yet, no matter how much money is poured into psychiatry, normalcy has proved to be unattainable by way of the money. If anything, statistics show that pouring more money was followed by a trend of an increased number of patients, no fewer suicides, and more medication prescribing.

As brain illness by definition is a dysfunction most of the time, it follows that normalcy is the opposite of dysfunction: normalcy is functioning. Thus, to reach normalcy, a system of reference is needed, to measure functioning against: functioning relative to what?

✳

Normalcy is relative to the patient. When navigating the patient's inner life, the psychiatrist has to be careful to maintain a balance between what to treat and what to preserve as normal for the patient. Take the example of a patient who dedicates ample time to a fascination with UFOs. The patient comes in asking for help with an unrelated problem, for instance spider-phobia. While tackling the preoccupation with UFOs is tempting, spider-phobia is what the patient came in for. The preoccupation with UFOs is not a concern, until the functioning of the patient is altered by it. Whether the preoccupation with UFOs is normal for the patient is not a matter of UFOs, but of how the patient functions while preoccupied with UFOs.

When the psychiatrist challenges a fascination with UFO's that is not altering the functioning of the patient, the psychiatrist risks treatment disengagement by the patient, which benefits neither the patient nor the psychiatrist. Thus, the psychiatrist must give weight to the patient's perspective on what is normal, even if it involves stretching the imagination of the psychiatrist into the realm of UFOs.

✸

Normalcy is not only relative to the patient, but is also relative to the psychiatrist. An illness to one psychiatrist is not an illness to another. Take the example of the controversy among psychiatrists memorized on the Internet on whether President Donald Trump experienced a psychiatric condition in the first few years in the office: some psychiatrists argued for it, others against.

✸

With normalcy being relative to both the patient and the psychiatrist, is normalcy ever absolute?

Let's take the example of an able person meeting a person with disability. The able person judging the person with disability as abnormal is a rush to conclusions. When, instead, the able person creates a mutually accepted shared reality together with the disabled person, this becomes their absolute normal. Likewise, say a hypomanic person meets a person with anxiety. When the two don't judge each other, the mutually accepted shared reality becomes their absolute normal.

The new normal is then not merely relative to each of the two people in the mutually accepted shared reality. Instead, it is absolute in togetherness. What was a relative normal from the point of view of the brain becomes an absolute normal for the mind.

✷

Whether relative to a brain or absolute to a mind, the normal is shaped by ideas. Thus, in general normalcy is a function of ideas. As ideas are dynamic, normalcy becomes dynamic. Take, for instance, a peaceful country led by the normalcy of peace. Despite being at peace, infusing the country with ideas of war on behalf of a greater good changes the state of normalcy, from peace harboring to a pre-emptive war strike.

✷

Ideas consistent with the absolute normal between two brains strengthen their mind. Ideas inconsistent with the absolute normal between two brains stretch their mind—until no common ground exists. When the mind dies, the absolute normal splits into two relative normals—one for each of the two brains that once formed the mind.

✷

To assume that normalcy is within the control of an average person without brain illness is tempting. However, ideas can be more powerful than people.

For instance, millions of people embraced the destructive doctrines of Nazism and Communism, not because they were inherently abnormal people, but because they were genuinely swung by powerful ideas. The tsunami wave of Nazism and, respectively, Communism, swung people in the twentieth century from normal into the abnormal. Likewise, in the twenty-first century, the wave of terrorism swings quite a number of people from normal into the abnormal, whether as single actors or part of an organized network.

For the average person without brain illness, the sense of normalcy can drift under the influence of ideas. For the average person with brain illness, the sense of normalcy can drift not only under the influence of ideas, but also under the influence of the illness.

Because ideas can be more powerful than people, ideas can control an individual, with or without brain illness. When doing so, they become non-autonomous ideas—not within the control of the individual, who is now lost to ideas.

Loss of control by an individual can happen to ideas and to illness. As for functioning, for illness to be present, a dysfunction is usually required. On the other hand, non-autonomous ideas can be present while still functioning. Take for example some politicians that are so ingrained with their ideas that not the politicians but the ideas are in control. These ideas become non-autonomous, outside self-control, but the politicians still function—and thus they are regarded as normal, despite being controlled by ideas.

When the controlling ideas are desirable, the loss of control to ideas may go unnoticed. But when the controlling ideas are undesirable, there is a need to regain control over ideas. Replacing the source of undesirable ideas with a source of desirable ideas becomes necessary. What makes ideas desirable is how ideas relate to the normal as defined outside the individual—the normal of the society. Infusing the normal of the society into the normal of the individual is done by a flow of desirable ideas from the society's perspective, against the source of the undesirable ideas controlling the individual.

The hope is that the flow of the desirable ideas from the society becomes a source of desirable ideas for the individual. The purpose of the society is for the individual to regain control over ideas, which in reality may turn out to be just the society regaining control over the individual. Even so, this can be a good step for an out-of-control individual, on the way back to normalcy. The social contract, the fabric that ties together people in the society, demands social normalcy over individual normalcy.

Through a flow of ideas, the society influences the normalcy for a person. For instance, when the society hires a psychiatrist to take care of a patient, the normalcy for the patient is fluid, subject to the flow of ideas between the psychiatrist and the patient. Consequently, whether the psychiatrist engages in a dialogue with the patient or remains silent influences what is normal for the patient.

ILLNESS

Ideas do not depend on the brain structure, as they are stored in a variety of ways over time; for instance, an idea may be discovered today written on a piece of paper a hundred years ago. Ideas flow through the brain. The relation between ideas and the brain can be compared with the relation between airwaves and a radio apparatus.

Among issues with the relation between ideas and the brain are:

1. *Multiple ideas overlap in the brain,* like several airwaves playing at once on a radio apparatus. The manic person has multiple thoughts, loudly bombarding the brain at the same time, like

a disco club with three floors on top of each other and no sound barrier in between;

2. *An idea is stuck in the brain*, like the same sentence being repeated over and over on the radio apparatus. The idea recirculates within the brain, and the brain cannot kick it out. This is common in patients with obsessive-compulsive disorder, and in patients with chronic auditory hallucinations;

3. *A new idea cannot be received by the brain*, like a radio apparatus missing a frequency band. Delusions in schizophrenia stop the brain from accepting new ideas, as they cannot be challenged rationally;

4. *Searching for a new idea by the brain cannot happen,* like a radio apparatus with the search button simply fallen off. In major depression without psychosis, searching for a new idea can become impossible, as illustrated by the lingering in the brain of an idea otherwise unwanted, such as "Life is sad," with the brain being too weak to move on from it.

Note that, even though ideas are not a function of brain structure, what ideas *stick around* in the brain is impacted by the brain structure. The flow of ideas is one thing. How the brain manipulates ideas is another thing.

Tension exists between, on one side, the flow of the ideas through the brain, and on the other side, the propensity of the brain to retain ideas depending on the structure of the brain. Two opposite forces are at play: the strength of the ideas vs. the propensity of the brain to retain ideas. Who wins depends on the difference in power between the two: ideas versus brain.

Consequently, when undesirable ideas are retained by an ill brain, desirable ideas become important to healing. Desirable ideas can become a flow that gets undesirable ideas unstuck from the brain.

✹

From a biological vantage point, brain illness represents a failure of nature to support the biological potential of an individual. From a holistic point of view, brain illness is a formidable roadblock on the creative journey toward fulfillment.

What exactly is, then, brain illness?

The scientific establishment looks at brain illness to be like a switch that accidentally got turned off in the brain. A great deal of evidence backs this up. The psychiatrist works to find a way, within the structure of the brain, to turn the switch back on. For example, the switch can be a chemical imbalance, needing rebalancing with medication.

A universe in itself, the brain illness is powerful, tempting, alluring, and strong. It's an attractor. The patient has to negotiate a way out of it. The bidimensional battle of good and evil in the healthy brain becomes the tridimensional battle of good, evil and disease in the ill brain.

Brain illness is the inability of the brain to efficiently manipulate information coming to the brain. It can originate in the brain. But it can also originate in the incoming information to a healthy brain, when the incoming information is detrimental enough to make the brain ill—that's when brain illness becomes a byproduct of what happens outside of the brain.

Likewise, healing requires the brain to regain the ability to efficiently manipulate information. This can be a brain-centered fix, for instance medication treatment to enhance the ability of the brain to manipulate information in order to unstuck undesirable ideas from the brain. Or, it can be an information-centered fix, like providing a stream of desirable ideas for the brain to float in, and thus to require less manipulation of information by the brain floating now in a desirable information. As ideas can be more powerful than people, a less than great brain in a stream of desirable ideas can do better than a great brain in a stream of undesirable ideas.

We know by now that the individual search for normalcy is influenced by what ideas are desirable, which, in turn, is determined by how normalcy is defined outside the individual. For instance, when a dysfunctional work environment requires, in order to get things done, a persistently angry boss, then the dysfunctional work environment validates as normal the persistent anger of the boss. This can preclude the angry boss to search further for normalcy. Instead, the angry boss may believe that normalcy has been reached already—with praises for getting things done, no less—when the persistent anger is desirable from the point of view of the work environment.

As brain illness can hide behind the way normalcy is defined outside the individual, a hidden brain illness is usually absent. This is because the individual *functions* relative to the normalcy defined outside the individual, and *dysfunction* is an element usually required by definition in brain illness. Note that the brain remains ill, continuing to have an inability to efficiently manipulate information. But despite the ill brain, the inability to efficiently manipulate information is masked by the absence of brain illness due to functioning relative to the normalcy defined outside the individual.

Note that just because the persistently angry boss *functions* in the dysfunctional work environment does necessarily take away from the same individual's *dysfunction* outside of the dysfunctional work environment, where the persistent anger is detrimental. Thus, brain illness stands the chance to go unmasked with a change in the environment, which can change how normalcy is defined outside the individual.

DIVERSION

In contrast with the normalcy defined outside the individual and *hiding* brain illness (when the normalcy defined outside the individual is aligned with the dysfunction of the brain, making the ill brain paradoxically function, and thus making brain illness absent), the normalcy defined outside the individual can also *cause* brain illness (when the normalcy defined outside the individual is misaligned with the brain function).

The causation paves the way for psychiatry to become a "dumping ground" for what happens outside the individual, turning psychiatry into a one-size-fits-all attempt at fixing larger issues, for which the psychiatric illness is merely a byproduct. Examples:

1. When the economy goes down, the stress goes up. People with no psychiatric problems before the economic downturn become patients now. They lose jobs, houses, spouses. Then, they show up to the psychiatrist, hoping to fix the psychiatric manifestations of the larger economic problems;

2. In families that do not give enough importance to togetherness, arguments knock at the door. The arguments can lead to depression and impulsivity. Then people show up to the psychiatrist, hoping to fix the psychiatric manifestations of the larger social problems;

3. Uneducated people miss an opportunity to learn values through formal education. Without values, temptations tend to become irresistible. Alcohol and drug use can follow. Then people show up to the psychiatrist, hoping to fix the psychiatric manifestations of the larger educational problems;

4. The decision to go to war is political, not psychiatric. However, war is a fertile ground for brains to become ill. War can break the brains of those who can no longer bear the burdens of fighting, despite having tolerated peace just fine. In times

of war, soldiers vulnerable to exposure to trauma develop post-traumatic stress disorder, or PTSD. Then soldiers show up to the psychiatrist, hoping to fix the psychiatric manifestations of the larger political problems.

This last example got to a new depth over time. A suggestion was made that the preventive use of an antidepressant can stop PTSD from happening. This means putting psychiatry not only at the back end of a larger problem of a non-psychiatric nature—war—but at the front end too. So, the "dumping ground" of psychiatry would not only be used reactively, when psychiatric manifestations of non-psychiatric problems happen, but preemptively too—just in case.

Taking care of the non-psychiatric problems, however, would make more sense than spearheading them with a psychiatric treatment in the absence of a diagnosis.

In general, psychiatry can become a diversion from focusing on the non-psychiatric problems that have psychiatric manifestations. Tackling the non-psychiatric problems by contributing parties, such as economists, sociologists, teachers, and politicians, can prevent the development of the psychiatric manifestations of the non-psychiatric problems.

The field of psychiatry is not capable to miraculously find a solution to non-psychiatric matters. Their "treatment" is outside of psychiatry.

RESPONSIBILITY

Besides not being able to resolve the non-psychiatric problems that have psychiatric manifestations, the field of psychiatry has another limitation: applying statistical knowledge to an individual patient. The psychiatrist cannot "do the work" for the patient, like a surgeon does. Instead, the psychiatrist provides a mere statistical approach to an individualized matter, which at times is beyond the reach of statistics.

The psychiatrist relies on preexistent knowledge about what generally works in a situation similar to that of the patient. The preexistent knowledge, being statistical, is not a guarantee for every individual patient. Searching for an individualized solution to a life's situation is for the patient to do, and cannot be replaced by delegating to the statistics through the eyes of the psychiatrist.

As the psychiatrist is not in the position to find an individualized solution for every life situation of every patient, the psychiatrist turns to the payor source for guidance: what does the payor source want in exchange for the money paid? Many times, the payor source is a governmental agency, something like the Department of Mental Health. For simplicity, let's call it "the state" (at a minimum, the state is in the position to influence the actual payor source by legal mechanisms).

The rules and regulations of the state tend to dampen down the individual variability of the patient through overregulation, which can end up chocking individuality. The rules and regulations of the state put a tension on the mind formed between the patient and the psychiatrist. The patient's brain either preserves the mind formed with the psychiatrist, by softening down from individuality, or disconnects from the mind, by springing into individuality.

When the patient's brain preserves the mind formed with the psychiatrist, by softening down from individuality, the patient faces losing the inner strength needed alone in front of private adversities—like an extended grief after the death of a close relative. The delegating of the individual responsibility to the mind formed with the psychiatrist can create an inefficiency for the patient by loss of the inner strength.

When the patient's brain disconnects from the mind formed with the psychiatrist, by springing into individuality, the patient is looking at dropping from treatment, remaining alone in front of the illness.

An example is a patient who only under persuasion agrees with the recommended treatment, then becomes non-adherent due to a concern with risks over benefits. The ignoring of the individual responsibility to engage in treatment can create an inefficiency for the patient by loss of treatment adherence.

Bottom line: the individual responsibility of the patient cannot be delegated, nor ignored, without facing a risk of inefficiency for the patient. To mitigate the risk, whether resulting from a loss of inner strength or loss of treatment adherence, the state has to support beyond rules and regulations an individual responsibility of the patient. If the state doesn't, the psychiatrist is called upon to go against the state's rules and regulations, siding with the patient instead.

✳

This brings us to the fact that the individual narrative of a patient is dampened by brain illness, leaving a void in the patient's core need to belong. The void becomes a fertile ground for the state to insert its own narrative, congruent with the state's values, against the individual narrative of the patient, congruent with the patient's values.

The state does the insertion of its own narrative by telling the story of how to solve the problems of people with brain illness: by tackling brain illness. But the state ignores that these are problems of people, not merely of brain illness—with solutions larger than treatment of brain illness alone. In effect, by becoming a story in itself, the state steals the show from the individual stories of patients.

Consequently, the state offers psychiatric treatment as solution, instead of supporting a diversity of individual solutions. Note that individual solutions have the right to exist, to push back against the state's attempt to replace the individual stories with the state's own story.

Granted, there comes a time when, for instance, chemical rebalancing through medication is necessary, by being the only intervention that can make a difference. But, to paraphrase an old saying, when hammering pays better, everything starts looking like a nail. Similarly, when the pay through the state is skewed toward prescribing medication, the state invites psychiatrists to prescribe.

On the question from the patient "What am I going to do?" the state becomes an active player when giving a financial incentive for an impersonal pharmacological intervention, like an antidepressant prescription. The financial incentive to prescribe competes against the need to address the source of what is happening with the patient—like isolation, lack of enough social support, and nothing to do during the day. When the source is the culprit, the chemical imbalance is the effect. But through focus shifting by giving a financial incentive to prescribe, the state inadvertently promotes the story that the chemical imbalance is the culprit.

The need to address the source of what is happening with the patient remains in place—for example, the need to replace isolation with a more thriving environment. Note that what constitutes a more thriving environment is not necessarily consistent with the state's narrative for the patient, based on the state's values, but is consistent with the patient's own narrative, based on the patient's values.

✻

This brings us to the topic of the separation of the church and state— the church being an example of an environment consistent with the patient's values, but not necessarily with the state's values (not all states encourage church). Within a patient's brain, the church and the state face off each other.

When the patient values the church and is treated by the state, can the state simply separate itself from the church inside the patient's brain? Is the separation even reasonable to try?

In reality, often the state finds itself in the surprising position of attempting to take over the role of the church inside the patient's brain—as if the state is turning into a new church. If it looks like a duck, swims like a duck, and quacks like a duck, it is probably a duck. The state preaches values to its constituents, by cleverly replacing concepts such as "church," "God," and "togetherness" with "government," "law," and "chemical imbalance." In doing so, the state competes with God inside the patient's brain.

For instance, the expansion of the definition of brain illness to trivial matters paves the way for leaving behind the importance of togetherness in solving life problems. Instead, chemical imbalance becomes a tempting excuse for prescribing medication for trivial matters that do not demand medication.

GOTCHA

From time to time, the professional literature says that a surprisingly large percentage of people have brain illness. Is that true? It depends on how brain illness is defined. Two erroneous ways to expand the definition of brain illness are described below.

The first erroneous way to expand the definition of brain illness is to look at what goes on at one point in time only, and to skip the need for *persistence in time* of what goes on in order to qualify for brain illness. Case in point: anxiety manifestations are normal after exposure to trauma. When skipping the need for persistence in time, just about anybody exposed to trauma can be "diagnosed" with PTSD.

The second erroneous way to expand the definition of brain illness is to ignore the usual need for *interference with functioning* of what goes on in order to qualify for brain illness. Case in point: a patient without cancer says "I am worried I might have cancer". That alone, without interference with functioning, is not the brain illness of hypochondria.

The over-diagnosis of brain illness by skipping criteria required for diagnosis, such as the persistence in time and the interference with functioning, leads to statistical distortions, unnecessary treatment, and funding misapplications, by deeming matters psychiatric for no good reason, instead of problem-solving outside psychiatry.

✴

Imperfections in behavior do not always belong to the realm of psychiatry. For instance, counting cars on the street, when mild enough to allow functioning otherwise, may be resolved by taking a road with less traffic, not by taking psychiatric medication. Or, fear of the dark may be resolved by placing a small lamp by the bedside, instead of going to psychotherapy. Or, insomnia may be countered by avoiding daytime naps, instead of undergoing an expensive sleep study in an overnight lab. Or, situational anger may be resolved by addressing the situation, instead of targeting the anger while not paying attention to the situation. In other words, imperfections in behavior can sometimes be addressed efficiently outside of psychiatry.

✴

Similar to how psychiatry blurs the line between imperfections in behavior and illness, so does the penal system capture imperfections in behavior under the umbrella of crime, by using an expanded definition of crime. For instance, in Missouri driving six miles over the speed limit is considered a criminal offense. Are then people who drive six miles over the speed limit criminals? People who drive six miles over the speed limit *are* criminals, according to the Missouri law.

But, given that almost all drivers in Missouri have driven at some point six miles over the speed limit, in effect the law regards the vast majority of drivers as criminals. Then, how can one respect the law, when the law is disconnected from fair justice? By leveraging a plea down to a fine, the law becomes a generator of revenue for the government more than a tool to achieve fair justice.

It speaks to the larger issue of the imperfections in behavior being sanctioned by a culture of the legal system that expects people to behave like perfect machines. This is in tension with the lack of perfect control over behavior when brain illness is present. By retaliating against imperfections in behavior, the law is harsh against people with brain illness.

Following are some examples of imperfections in behavior. Raising the voice to someone in public could be called the tort of assault—a civil issue of law not requiring physical contact. Trying hard to explain one's view to an arresting officer could be easily interpreted as resisting arrest. Telling a woman one time that she is beautiful can be called sexual harassment. Asking a stranger for a spare swipe of a subway card can result in a charge of theft of subway services in New York City. Screaming at a drunk husband that gambled his paycheck away can be interpreted as the crime of disorderly conduct.

What's wrong with this picture? Nothing, according to the law. A lot, however, is troubling for people with brain illness. The people with brain illness are at risk of legal consequences for imperfections in behavior due to struggling to control behavior. But who truly has perfect control over behavior in all circumstances, no matter how stressful? A buffer for tolerating imperfections in behavior can ground the law in reality, not in utopia.

Unfortunately, in an isolationist culture that deals at arm's length with people, intolerance seems too often to be the answer.

✴

During an arrest, when giving the Miranda warnings, the police officer essentially says, "Don't talk to me, or you'll get convicted on your own words." If after the Miranda warnings the suspect still talks, the officer applies an instant "intelligence" test, assessing the suspect's capacity to make decisions. This so-called intelligence test almost guarantees that an *un*intelligent decision—to give details to the cops—will stand in court as confession when the officer deems the suspect intelligent enough to know better.

The confession paves the way for a conviction, which is why the "intelligence" test might as well be called a stupidity test. The ability of the suspect to weigh options is not enough to actually be intelligent, when the option selected is *not* intelligent. But, in the cat-and-mouse game of arrests, people making an unintelligent decision can be deemed intelligent, and the confession stands.

Note that, to be completely truthful, the cop would need to say more to the suspect than the currently used "you have the right to remain silent." The cop would need to unbundle verbally both the layperson's meaning of the "right to remain silent", and the hidden legal meaning of the "right to remain silent." The full statement by the cop would need to be: "You have the right to remain silent but I can continue to question you; and, you have the separate right to stop me from questioning by saying that you want to remain silent before actually remaining silent."

Hidden in "the right to remain silent" is the separate right to force the cop to stop questioning, activated only when the suspect does not remain silent, but speaks up first, by saying "I want to remain silent." Without first verbally invoking the right to remain silent, simply remaining silent allows the cop to interpret the silence as mere acquiescence to questioning, in effect allowing the cop to continue questioning.

To add insult to injury, there is even a subtler way to obtain a confession than by outsmarting the suspect with a hidden legal meaning underneath the words "you have the right to remain silent." The officer is allowed to outsmart the suspect by behavior alone. Playing a "good cop" projects an image of inherent trustworthiness, for instance when the cop offers a cup of water to the suspect. Already in need of support, the suspect might then initiate a conversation about the alleged crime, as if to unload the psychological burden on someone who is now trusted. By initiating a conversation, the suspect waives the right to remain silent.

Note that the waiving of the right to remain silent does not have to be verbally asserted ("I am waiving the right to remain silent"). Instead, the mere conduct of unsolicited confession is the implied waiver of the right to remain silent. The cop offering a glass of water is not considered solicitation, but can get a suspect to talk faster than by asking a barrage of questions. The officer who offers a glass of water does not necessarily intend to quench the suspect's thirst, but may intend for the suspect to develop trust in the "good" cop. The developing of trust then traps the suspect into a confession deemed unsolicited, which can stand in court.

In an adversarial legal system, the police officer is the legal adversary of the suspect. By hiding underneath seemingly plain words a different meaning to the words, and by hiding underneath seemingly plain behavior a different intent, the officer plays mind games, which can result in a confession by the suspect. While the use of physical force by an officer to coerce a confession is not allowed by due process, the use of mind games for a confession is common practice.

Due to the dysfunction required usually by definition in brain illness, people with brain illness can be particularly vulnerable to mind games played by cops.

*

On a daily basis, police officers are called to deal with people with brain illness and to decide whether the destination is a hospital or a jail.

In the hospital, the patient gets medical care. In jail, the person with brain illness goes to "the pen." Here is how the pen might look on a busy day. Thirteen inmates share the pen, waiting for the psychiatrist to show up. In the back, an inmate holds the toilet bowl with both hands. He repeatedly shoves his head into the bowl and pulls it out again, screaming every few seconds. Nearby, an inmate runs in place, as if on an invisible treadmill. A few inches away, two inmates jokingly box each other, throwing light punches. To the right, another inmate holds up his hands, while throwing his entire body toward the wall, repeatedly. He literally tries to jump through it. A few inmates sleep on the floor, as if having the quietest time of their lives. Another inmate repeats the same foul phrase made of two words (one starting with an F, the other with a Y). A couple of inmates dance furiously to a "Cotton-Eyed Joe" kind of song only they can hear. To the left, a few inmates mimic singing along, in silence. To top it all off, one inmate is in the process of pulling out his private part, as if to urinate on the Plexiglas that separates the inmates from jailers. When a jailer jumps up to stop him, the inmate puts it back in the pants. As the jailer retreats, the inmate attempts to pull it back out, to aggravate the jailer. That's the environment a person with brain illness can land in, when brought to the pen of a jail, instead of a hospital.

The police officers have the power to send a person with brain illness to a supportive environment, conducive of recovery (the hospital), or to a challenging environment, at risk of more disruption (the pen of a jail). Besides the nature of the alleged crime committed, the police officers base their decision on the severity of the psychiatric issue—traditionally not the cops' strongest suit. Training police officers on brain illnesses becomes a necessity, not a caprice.

✳

A reason many psychiatric patients face cops on a routine basis is a discharge too soon from the hospital. The discharge is done in the name of freedom of movement, under pressure from health insurances—who decide for themselves for how many days to pay for the hospital stay—and under pressure from hospital administrators—who don't like to have days unpaid at the end of the hospitalization. Here is an example of how it all plays out.

The patient comes in the hospital with a clinical condition needing fourteen days to recovery. On day five, the patient feels better. Far from being recovered, the patient demands to leave the hospital on day five. The health insurance regards the feeling better by the patient as a reason to stop payment for the hospitalization on day five. The hospital administration uses the utilization review department to press the doctor for discharging the patient when the payment by the health insurance stops on day five.

The law gives the final blow: because the patient does not appear a danger to self or others on day five as observed in the hospital environment, the patient has the right to walk away, and the doctor cannot do much about it.

This brings us to the engineering discipline called Strength of Materials, dealing with the ability of any material, for instance steel, to withstand without snapping a pressure, such as a weight. To determine the resistance, the material is tested under a pressure. Just because the material does not snap today, in the absence of a pressure, does not mean the material will not snap tomorrow, under a pressure. Likewise, the patient in the hospital may look "fine" on day five under no pressure, but the real test is to estimate the behavior of the "material" in the foreseeable future, under the weight of a stress factor present only outside of the hospital.

In psychiatry, the stress factor can generally be conceptualized as an obstacle for the patient to deal with. Obstacles can be many outside the hospital, depending on the circumstances of the patient. They need to be sorted out in order to anticipate accurately the behavior of the "material" in the foreseeable future.

For practical reasons, who else than the psychiatrist is in the best position to estimate the snapping of the "material" under the weight of a stress factor in the foreseeable future? After all, this can be a clinical snapping, not necessarily within the patient's control.

We are on shifting sands. In order to prevent a foreseeable snapping from happening, the psychiatrist may need to convince a judge to restrict the freedom of the patient. This requires the letter of the law to be on the side of the psychiatrist. But the letter of the law can be vague. Plus, to merely get to see the judge, the psychiatrist needs the logistic support of the hospital administration. In turn, the hospital administration is not very thrilled to offer logistic support when the health insurance stopped payment because the patient looked "fine" on day five in the stress-free environment of the hospital.

After the patient is released too soon from the hospital based on looking "fine" in the stress-free environment of the hospital, the patient is not ready to handle the level of stress usually present outside the hospital. Consequently, lost in the newly found freedom, the patient may develop a crisis, and the community might call the cops on the patient.

Liberty from others is of great value to many people. Wars have been fought in the name of it. Once liberty from others is attained, the next question is: will it hold water? Brain illness is not another culture, another ethnic group, or a constitutionally protected right. Instead, brain illness is usually a vulnerability of not functioning.

It was once said that too much freedom results in loss of freedom. For example, drawing a house on paper can be done freely, but to withstand gravity, building a house must follow construction principles. Without discipline when building, the house caves in. It takes self-discipline to create beauty.

Like a house needs to be built by following construction principles in order to withstand gravity, treatment in psychiatry needs to follow treatment principles in order to withstand brain illness. One principle is that a psychiatric hospitalization cannot be too short compared to what is needed to achieve recovery. The length of the required hospitalization must be respected, or else brain illness wins against an inefficient psychiatric treatment. Despite the wishful thinking of some extreme advocates, actual freedom is not automatically achieved by cutting short the length of the psychiatric hospitalization necessary to recover. Instead, patients face a worsening of the psychiatric illness when released from the hospital too soon.

The real confinement is, first and foremost, inside the patient's damaged brain, where the psychiatric hospital plays the role of a bridge to freedom. Shattering too soon the walls of the psychiatric hospital damages the bridge to freedom. Without it, the patient is looking at remaining brain-confined within the illness.

Moreover, when a patient acts seriously and persistently disturbing after a release too soon from the hospital, a new confinement in either a hospital or a jail is brought back to the table on behalf of the public interest by the cops. When a too short hospitalization leads to the patient not having enough self-control over the illness, the remaining free in the community is in peril, because the ability of the patient to engage in a balancing act between freedom and self-restraint is jeopardized.

Note that in the hospital the lack of self-control by the patient over the non-autonomous, automatic process of illness is acknowledged.

But if jail turns out to be the next stop after being released from the hospital too soon, the presumption in jail is that the individual is autonomous, in self-control. Consequently, the jail deprives the ill individual of the necessary focus to regain enough self-control against the illness. Instead, the jail risks to confine the patient even more within the ill brain.

✳

Going to jail invites the prosecutor to the table. In general, dealing with a prosecutor seems on surface to be a clear-cut matter: you either did it, or you didn't do it. For instance, if Tom steals Sam's duck, then Tom is a thief. But, this requires a closer look. Was it stealing? One of the first things the prosecutor explores is the *mental state* of the defendant at the time of the alleged crime. It's not just about the facts. The law transgresses into the realm of psychology, where the mental state is a controlling factor in guilt assignment.

The four typical culpable mental states in law are: *intentionally*, *knowingly*, *recklessly*, and *criminally negligent*. The question becomes: What was in Tom's head at the time of the alleged crime? In other words:

1. *Intentionally:* Was Tom's objective to take the duck from Sam?
2. *Knowingly:* Was Tom aware that the duck was going to come his way when he opened the door?
3. *Recklessly:* Was Tom careless in opening the door, by ignoring the risk seen by Tom that the duck, in Tom's sight at the time, might come his way?
4. *Criminally negligent:* Should Tom have anticipated the risk not seen by Tom when opening the door, that the duck, not in Tom's sight at the time, might come his way?

In all four examples, the manifest action is the same: the duck belongs to Sam and is found in Tom's yard. With Sam claiming that the duck was stolen, what makes the difference between Tom being guilty and not guilty is what was in Tom's head at the time of the alleged crime—because a culpable mental state must be proven in order to be found guilty. In the absence of a culpable mental state, Tom is not guilty. The prosecutor has the burden of proving the culpable mental state of Tom.

In addition to being either guilty or not guilty, the law creates a third category: not guilty by reason of insanity. Does this mean, if Tom has a brain illness, then he is legally insane? Not so fast. Insanity is an example where the law divorces the technical legal meaning of a word from the plain meaning of the same word. To prove insanity, proving brain illness is not satisfactory in itself, and proof beyond brain illness is necessary. The required proof beyond brain illness comes in various forms, but usually revolves around an inability of the defendant to appreciate what the defendant was doing, or an inability of the defendant to act right.

Without an insanity defense, the law presumes the defendant to be sane. The prosecutor is then blind to the brain illness of the defendant, despite the brain illness impacting the mental state (that the *brain* illness impacts the *mental* state is a no-brainer). To show insanity, the burden of proof falls on the defendant: legal insanity is an affirmative defense, meaning the defendant has first to acknowledge the crime—"Yes, I did it"—then to justify it—"but I was insane at the time of the crime." After introducing the insanity defense, the presumption about the defendant switches from innocent until proven otherwise, to guilty until proven otherwise.

A person with brain illness who tries and fails to prove insanity falls into the trap of affirming "Yes, I did it," a prerequisite to use the insanity defense. Now the defendant is sane for good—presumed first, then established by failing to prove insanity—and, all of a sudden, the defendant is very close to getting a guilty verdict—because of the affirmation "Yes, I did it" made at the time of introducing the insanity defense. The affirmation stays in place despite failing to prove insanity.

That's why plenty of people with brain illness choose to avoid the insanity defense—where the prosecutor has to prove nothing—and play sane instead: to put the burden of proof on the prosecutor.

Another reason why many people with brain illness choose to avoid the insanity defense, and play sane instead, is the following: when playing sane, the duration of the stay in prison is set by the time the prison starts, depending on what the defendant did *before* the verdict. When playing insane, however, the duration of the stay in the forensic hospital is not set by the time the hospitalization starts, depending on what the defendant does *after* the verdict—in the hospital.

Thus, the mechanics of justice do not favor bringing to the table the issue of insanity, given that the use the insanity defense is risky both before the verdict (the defendant's failure to prove insanity turns into the admission "Yes, I did it") and after the verdict (a not guilty by reason of insanity verdict turns into a stay in the forensic hospital with a duration unknown upfront).

So, arguing "madness" is not a smooth route. Well, that unless the charge is murder and the right set of circumstances is met which would be maddening to a reasonable person: a situation indicative of *heated passion*. This kind of right set of circumstances opens the door to the affirmative defense called extreme emotional disturbance, used to avoid being found guilty of murder despite the defendant acknowledging "Yes, I did it."

The affirmative defense of extreme emotional disturbance does not require the defendant to prove brain illness, nor heated passion, but only to justify the heated passion by a mere reasonable explanation or excuse, in the right set of circumstances (e.g. "my spouse was making out with the victim, which was a situation indicative of heated passion that would be maddening to a reasonable person; being upset at the sudden realization that my marriage was taken away from me,

I've acted the way I acted, and the victim is now dead"). Brain illness does not have to be mentioned at all, much less proven. It's not brain illness that creates the madness then, but the right set of circumstances—or so they say in the penal system.

PERSONAL

Being the victim of a crime is personal to the victim. On the other hand, when suffering from a disease, how much more personal can it get?

Sure enough, the disease feels personal to the sufferer. But the disease does not belong to one person. It replicates across the board, with a common core despite marginal variations. The disease becomes impersonal, despite feeling personal. Thus, when a psychiatrist looks at sick people, it makes sense to employ an impersonal approach to tackling disease, similar to fixing a broken leg. Using an example from psychiatry, ten milligrams of an antidepressant is the same ten milligrams for everybody needing it.

Now, just because the disease becomes impersonal, does not mean the treatment has to always remain impersonal. For instance, a mildly intellectually disabled individual that gets emotionally agitated due to a lack of cognitive skills to control emotions can be treated two ways.

One way is to create a generic solution, by using a mood stabilizer, like Depakote, to treat the symptom of mood disturbance.

Another way is to understand better what gets the patient frustrated from the environment, then to adapt the environment to the ability of the patient to tolerate it emotionally. Chances are, the patient will improve. Why? Because the mood does not need a mood stabilizer when the limited cognitive abilities due to mild intellectual disability allow a *normal* raw mood variation to come through unfiltered—needing instead an adapted environment. Here, the cognitive filter that would regulate a normal raw mood is inefficient, giving the impression of an unstable mood. The culprit is the cognitive filter, not the mood. A minor frustration, which would otherwise be softened down cognitively, ignites the normal raw mood, unfiltered cognitively due to the mild intellectual disability.

Psychiatry is an attempt to use science on personal matters. This reaches a limit: the psychiatric studies look at patients as numbers, centered on the brain, and relying often on the chemical imbalance hypothesis. The psychiatric studies compete with the personal touch of those who are not psychiatrists but form a mind with the patient. Through the mind, those that are not psychiatrists facilitate healing by lending to the patient the power to reason *together* through the lens of trust, personally touching the impersonal disease. In the example of the patient with mild intellectual disability that gets emotionally agitated, reasoning together by the power of the mind formed with a supportive caretaker helps the ill brain soften down cognitively an otherwise normal raw mood.

✳

Because psychiatry deals with personal matters, it is not hard to inadvertently upset a patient. Whatever is being said can quickly turn personal to the patient, who can then claim being offended. What is neutral from the point of view of the psychiatrist can be hurtful to the patient, for instance because of a previous experience that the patient did not yet bring to the attention of the psychiatrist.

In general, offensiveness is a state of mind, made of two brains, depending not only on the speaker, but on the listener too. Whether a seemingly neutral statement is offensive depends on both what is said and how it is heard. The speaker does not always have time to screen for alternate meanings to an innocent meaning. The listener that presumes an innocent meaning fosters dialogue, and thus the formation of the mind between the two brains, of the listener and of the speaker. On the other hand, the listener that is on the lookout for an offensive meaning favors silence by the speaker, and subsequently the death of the mind once formed between the listener and the speaker.

Because patients can be hurt easily by seemingly neutral statements, the psychiatrist becomes tempted to give up on much of the dialogue. No wonder plenty of psychiatrists choose to stay mostly silent. In general, the field of psychiatry attempts to shrink itself into finding commonly acceptable language that does not offend anyone. Consequently, psychiatry finds itself in the corner of almost silence, where medication prescribing reigns.

Taking the opposite approach, of a dialogue on challenging topics, is an opportunity for the psychiatrist to enrich the experience of the psychiatric visit. But courage is necessary for the psychiatrist to go beyond mostly listening, to speak up despite the risk of being perceived as "offensive" due to a hidden meaning identified in hindsight by the patient through borrowing content out of context.

✳

One of the challenging topics of dialogue that risks a pushback on offensiveness ground is to discuss the personal beliefs of the patient. But psychiatry intertwines with the inner layers of the soul, where personal beliefs are, so looking at the personal beliefs of the patient comes naturally to the realm of psychiatry.

The psychiatrist-patient relationship is influenced by the belief system of the patient, as well as by the belief system of the psychiatrist—despite the psychiatrist trying to hide behind a scientific persona. The influence of the belief system of the psychiatrist on the psychiatrist-patient relationship depends on how freedom from bias is defined.

Freedom from bias can be defined on a range, from taking distance from the belief system of the patient, to aligning with the belief system of the patient. When freedom from bias is defined as taking distance from the belief system of the patient, anything less than the distance is a bias. When freedom from bias is defined as aligning with the belief system of the patient, anything less than the aligning is a bias.

At a minimum, to be free from bias, the psychiatrist needs to take into account the belief system of the patient—otherwise the psychiatrist would be biased against the personal nature of the soul examined.

＊

What works better: pairing the patient with a psychiatrist with the same belief system, or with a psychiatrist with a different belief system? Let's take the example of the most common belief system: religion.

When the patient and the psychiatrist are of the same religion, one would assume the treatment works better because of the mutual understanding between the patient and the psychiatrist through a shared belief system. But, brain illness may twist the meaning of religious concepts, leading to different meanings for the patient than the rest of the people of the same religion. Brain illness may color the meaning, pushing it far from the norm. When because of being of the same religion, the psychiatrist does not verbally explore the religious concepts with the patient, the psychiatrist can miss the significance assigned due to brain illness by the patient to religious concepts.

When the religion of the psychiatrist is different than the religion of the patient, the difference in belief systems creates an opportunity to explore the spiritual concepts of the patient in words, without assumptions. But the conversation can get awkward, because of an inherent hesitancy of people to discuss spiritual concepts when religions are different.

In general, whether of the same belief system or not, the psychiatrist is called to immerse in the patient's perspective, by adopting a student-like role. The immersion begins by the psychiatrist eliciting the patient's own perspective, in order to take in account what the patient means, not what the psychiatrist assumes that the patient means.

SUBJECTIVE

Patients talk about all sorts of things in therapy. Some of what is talked about does not make sense objectively. Is this still important? In practical life, what does not make sense objectively is not important—because it doesn't add up to a practical application. But in the inner life, what does not make sense objectively can still make sense subjectively.

Here are examples of what patients might say that does not make sense objectively, and still makes sense subjectively, for the patient: "I met with the heirs of Elvis Presley and we are opening a million-dollar music shop around the corner"; or, "I've got a beautiful house, a great job, a wonderful spouse, but I hate my life"; or, "The government put a chip in my head to spy on me"; or, "I'm terribly afraid that something wrong is going to happen to me, even though I don't know what." These examples make sense subjectively, for the patient, because they create real feelings in the patient, and because they represent the patient's perception of reality, on which the patient acts.

How can the psychiatrist become logically immersed in a patient's inner world that isn't based on logic? How to make sense of the shuttered inner world of the patient?

One can begin by learning it piece by piece in treatment sessions, with an attitude of nonjudgmental acceptance necessary to engage the patient in treatment. Then, in analyzing the odd puzzle unfolding, the psychiatrist can start with a piece of the puzzle where a change makes the most difference.

The subjective point of view of the patient is the objective reality that the psychiatrist has to work with.

*

I would be remiss if I don't bring up the need for self-protection by the psychiatrist. There's a saying that being a psychiatrist for too long can open the door to brain illness.

Even though the psychiatrist gains strength after years of training, the psychiatrist can "lose it" by continually having to validate non-functional points of view for the sake of keeping patients engaged in treatment. Constantly navigating through broken brains can stretch a psychiatrist's brain to the point of rupture.

Look at it this way: When a parade of fifteen people a day try to convince you that an abnormal thing is normal, at some point you wonder, "What if they're right?" That's exactly when the psychiatrist needs to stop, take distance, and get back in touch with reality. For instance, merely because fifteen patients say a Martian is in the basement does not mean they are right.

Despite trying to project a strong persona, the psychiatrist is vulnerable, having to balance not only patients, but oneself too.

*

All day long the psychiatrist deals with patients' losses, struggles, despair. The psychiatrist listens to stories about broken relations.

After the roller coaster of hearing a variety of broken stories for long hours, the head of the psychiatrist can be spinning. At the end of the day, the psychiatrist may need a break from everything— no movies, no radio, no books. No more thinking. Simply let the silence sink in.

Or, symphony music may work well for the psychiatrist; merely listening to it involves no overt story, no explicit characters to follow, no epic battles, no plots, no logic. Rather, it's a shower of sounds after a long day in the mineshaft.

Talking with patients with brain illness can feel like drilling in stone. The psychiatrist drills and drills, only to hear the sound of the stone at times echoing, "It's just a rock here, Doc. Give it up." Still, the psychiatrist hopes that, by continuing to work with the patient, one day the stone will spring back to life.

Being a psychiatrist may seem fulfilling because of taking care of others. But fulfillment is not easy when many patients have an abysmal cure rate. Plenty of psychiatric illnesses are chronic, long-term, without a cure in sight. And yet, the psychiatrist continues to try tearing down the barrier of brain illness to health.

POWER

An old psychoanalytical theory compares the patient with a toothpaste tube. The theory says that, in order to modify emotions, they have to be *squeezed out* from the patient by the psychiatrist in form of words.

Getting the patient to talk through emotions used to require eliminating all distractions, including the very sight of the psychiatrist— for the patient to be left reading emotions from inside. Hiding the psychiatrist behind the couch was *the* way to conduct treatment in the time of psychoanalysis.

Nowadays, the psychiatrist is visible in front of the patient. But because the patient still needs to look inside, the patient adapts, by looking away.

In a room with a blank board and a window, clinical observation has shown that patients tend to look through the window instead of at the blank board. The patients are unconsciously attracted to the power of the minds waiting to be formed beyond the window, as opposed to a brainless blank board.

✳

Earlier in the book we introduced the assumption that two brains working together are more powerful than the same two brains working simultaneously but disconnected.

Scientists have long considered how a mind differs from the brain. People don't live so much in the physical space of the brain, visible on a CT scan, as much as in the psychological space of the mind.

With a pair of brains having a mind of its own, the question is which brain influences which: Does the healthy influence the ill? The ill influence the healthy? The good influence the bad? The bad influence the good?

In reality, the two brains are in a creative tension within the mind. A chain of power forms: the more powerful brain uses the mind to influence the less powerful brain. This can be an advantage, when good ideas—that otherwise would not get to the less powerful brain—can find their way from the more powerful brain, through the mind, to the less powerful brain. But it also creates a potentially dangerous situation, when bad ideas—that would otherwise be rejected by the less powerful brain—can find their way from the more powerful brain, through the mind, to the less powerful brain.

It's a delusion of grandiosity that most brains are truly autonomous. Brains are under the influence of minds.

＊

Besides relying on itself for a source of power, the brain has a separate source of power: the mind. A powerless brain can still find a source of power in the mind.

An example of the power dynamic between a brain and the mind comes up when dealing with brain illness. The brain may lose its own power during brain illness. Consequently, keeping the mind alive during brain illness becomes of critical importance, to maintain an alternative source of power for the brain.

When the brain that lost its own power disconnects due to brain illness from the other brain, the last source of power for it is lost—the mind. However, with the mind being kept alive, the healthy brain of the mind can turn into a physical key for the ill brain, unlocking the potential of the ill brain. This is well depicted in the movie *A Beautiful Mind*, in the scene where the healthy wife convinces the ill husband to take his medication: the mind formed between the wife and husband gives power to the healthy brain of the wife to unlock the potential of the ill brain of the husband, for taking his medication, and thus for stepping forward to health. In doing so, the mind between the wife and the husband is, indeed, a beautiful mind.

To help the patient efficiently navigate through opportunities for improvement, the psychiatrist is called upon to connect with the patient within professional boundaries, through the mind. Power lies not only in the ill brain, but between brains too. The power of the mind touches the ill brain. The ability to heal resides not merely in the physical structure of the ill brain, but also in the mind.

Healing is not restricted to an individual sport. Healing can stem from the interaction between players, being a team game. The interaction between players can make the difference when an individual player cannot. Distinct from the power of the brain, the power of the mind is not to be neglected when dealing with brain illness.

Here is another example of the mind as a distinct source of power for an ill brain: *peer support* (the mind) facilitating the recovery of an alcoholic brain.

Note that the power of the mind can work in reverse, turning a healthy brain into being ill. For instance, *peer pressure* (the mind) can turn a healthy brain into being ill with alcoholism. Here, at a minimum, the mind is a risk factor for brain illness. Other examples of minds being a risk factor for brain illness are: arguments with a spouse after years of good marriage; the stressful pushback against an abusive supervisor; or, the subtle impact of a friend with inefficient values.

The psychiatrist is called upon to address the minds that are a risk factor for brain illness. Unfortunately, the psychiatrist is not well equipped to do so. Often, the psychiatrist is left with only talking about these minds with the patient. But the mere talk does not necessarily make the minds that are a risk factor for brain illness go away.

However, the conversation between the psychiatrist and the patient is an opportunity to pump the mental muscle in favor of the patient: the mind psychiatrist-patient can be a protective factor, to improve the prognosis of brain illness.

Unfortunately, psychiatrists can choose to stay silent, where no dynamic tension is created between the psychiatrist and the patient, having the practical effect of an absent psychiatrist, as far as the mind psychiatrist-patient is concerned. Then, the mental muscle does not pump (the mind is practically absent) and the ill brain repeats for the psychiatrist a monologue already known to the patient.

✳

As it stands today, psychiatry is over-focused on the treatment of the ill brain and under-focused on the power of the mind to rehabilitate the ill brain. Note that to rehabilitate the ill brain does not require returning to a symptoms-free state, impossible in many chronic brain illnesses. Instead, to rehabilitate the ill brain, a mere recovery of functioning is needed.

Usually when rehabilitation through recovery of functioning is achieved, brain illness goes away by definition, even with residual symptoms still present, because brain illness usually requires by definition a dysfunction separately from symptoms.

Case in point: a patient seems depressed after divorce. When a complementary intervention without medication achieves the rehabilitation of the ill brain by recovery of functioning, the psychiatrist does not have to start an antidepressant. For instance, when the patient moves in with a supportive sibling and achieves the rehabilitation of the ill brain by recovery of functioning, the need to medicate is replaced by the power of the mind between the patient and the supportive sibling, without medication. Said another way, the mind is the medication.

Note that in practical life achieving rehabilitation by moving in with a sibling does not automatically exclude medication, as the strength of the ill brain without medication may be significantly less than with medication.

✳

There is a difference between brain illness and an ill brain. By definition, brain illness usually involves an ill brain that is functionally decompensated. But just because brain illness disappears when the brain gains functional compensation (e.g. through the power of the mind), the brain does not automatically stop being ill. The brain may very well remain ill, but falling outside of the usual definition of brain illness due to being functionally compensated.

The operating concepts of ill brain versus brain illness need to be refined in the future.

Despite an ill brain having not yet evolved into a brain illness, strengthening a functionally compensated but ill brain is still needed. The mere foreseeability of brain illness demands the attention of the psychiatrist. Notwithstanding ethical concerns raising a barrier against medicating an ill brain without brain illness, the need to prevent brain illness remains in place. Functional strengthening to avoid functional decompensation becomes the preventive intervention of choice. What is left for the psychiatrist to do, when medication is not an option? Speaking up seems like a natural choice.

Even in severe illnesses, like schizophrenia, a strong support system can give functionality despite residual symptoms, leaving the brain illness as a mere incidental finding on a brain scan. The patient can be so functional that the psychiatrist cannot even tell at the first glance that schizophrenia is at play. That's because schizophrenia is not at play, despite residual symptoms being later identified as having been present.

Since brain-centered interventions, such as medication prescribing, cannot cure many brain illnesses, the mind is the limit in psychiatry. There lies the problem of an isolationist society, where each individual is treated at arm's length: the psychiatric problem is regarded as that of the patient, and not of anyone else's.

When the perspective changes, from "What's wrong with the patient?" to "How can I help the patient?", the door to recovery *together* opens, increasing the likelihood of recovery compared with brain-centered interventions alone. The power of the mind can do what the brain cannot.

The mind can pull the patient away from brain illness. For instance, the mind can guide the patient through psychiatric treatment, facilitate attending appointments, and encourage taking medication as prescribed. By achieving a functional compensation of the ill brain, the mind leaves the ill brain without brain illness. In doing so, mental health overcomes brain illness.

✳

Clinical observation has shown that some brains are toxic to the mind. Let's take the example of a brain loaded with negativity. The negativity brings toxicity into the relationship, making the mind bitter. Then, when spreading negativity to another person, the brain feels a temporary sense of relief. It's as if for a while the negativity has finally left the person and found another hiding place.

For protection against negativity, family members and friends are looking at ignoring the individual loaded with negativity, who slides into isolation, away from minds. This results in a loss of mental power. Subsequently, when the only other source of power is also lost— the brain power—the brain is left powerless. Note that the brain power can be lost to brain illness, or to sticking to inefficient values in absence of brain illness.

Sometimes the frustration of being powerless kills—by suicide, or by killing the other brain of the gone mind, or by killing both brains of the gone mind. For instance, from time to time news channels talk about a family in which both spouses just died of a suspected murder-suicide. It raises the possibility that the brain of the killer lost both types of power: the power of the brain and the mental power. An important source of power in the marriage was the mind between the spouses. When both powers are lost, of the brain and mental, the frustration of powerlessness can turn into the murder-suicide of the spouses.

It's worth mentioning here that, by far, not every person who loses both powers, of the brain and mental, resorts to murder-suicide. But, a common ingredient in suspected murder-suicide cases between spouses—or between employer-employee, or between other people who once formed a mind—seems to be the loss of both types of power, that of the mind and that of the brain, making the killing of the other person more than a random killing. The connection with the other person killed was once a mind.

Finding out what determined the death of the two persons in an apparent murder-suicide requires proceeding with caution. Like for a suicide, a medical analysis and a forensic analysis need to be completed, to find out the factors at play—medical and forensic. Despite the appearance, not every double death of people that once formed a mind is a murder-suicide. The details of what went on hold the key to what killed the two persons.

✳

Before completely losing brain power to brain illness, a patient may experience a serious decrease in brain power—in light of the dysfunction usually required by definition in brain illness. When the decrease in brain power becomes protracted, the patient may give up on trying to get better, missing new opportunities to regain the lost power.

Subsequently, to form a mind with the patient who gave up on trying to get better, the psychiatrist has the task to start a conversation. It comes up, for instance, on a medical unit of a hospital, after a suicide attempt by the patient. With hope eluding the patient, discovering hope together is the chance to bring back the patient to the lost power. When the patient is silent, the burden is on the psychiatrist to take the first step in the conversation.

Facing a patient unable to move forward in life is an opportunity for the psychiatrist to use the power of storytelling, to infuse power in the brain of the patient. This is in tension with the traditional role of the psychiatrist, which revolved around listening.

As "believe in yourself" does not work, "believe in a relationship" finds an opportunity to surface, being both a challenge and an opportunity. A challenge because the psychiatrist faces the temptation to portray oneself as the holder of the key to power. An opportunity because the psychiatrist can lead the patient from a decreased power of the brain to the increased power of the mind. The "believe in a relationship" opens the door to believing in the power of the story.

Carefulness is needed though. The switch from traditional concepts of reality to a virtual environment taps deeply into the unconscious of the patient, challenging the line between what is real and what is not, through a power play. A similar power play takes place in movie theaters: the audience, experiencing a loss of power in front of adversities on the screen, look for a character to fill the power void, such as Superman or Wonder Woman, Batman or Spiderwoman. Patients too, when losing power in the context of the dysfunction of brain illness, look for a character to bring the power back.

The Psychiatrist is a character that can give the illusion of power to the patient, in light of the patient's imagination. But at the end of the day, the psychiatrist is merely an actor playing the character of The Psychiatrist, struggling to bring the patient to a reality that withstands the test of time.

✳

When losing power due to brain illness, the brain gets hungry for power in order to avoid being stuck in a protracted decrease in power. In turn, the hunger for power is in the position to kick off a desperate search for an alternative source of power than the brain losing power. Of the minds that can be a source of power, feeling powerful in the relationship with the psychiatrist becomes tempting.

To make the patient feel powerful in the relationship with the psychiatrist, the profession of psychiatry is centered on patient satisfaction, validating what is pleasant to hear for the patient. Note that many patients like to hear an echo by the psychiatrist of their own opinions, already formed by the time of stepping into the psychiatrist's office.

Not validating the patient's opinions is risky business, as the patient can simply disengage from treatment. Tactfulness is necessary. For instance, a patient might complain that raising a child is stressful, then asks for Adderall and Xanax. When the psychiatrist says "You are doing a great job in raising your child, you do not need Adderall or Xanax," the treatment is at serious risk of ending.

The psychiatrist might want to begin by avoiding confrontation, because the confrontation can make the patient go through a loss of power. Instead, a non-confrontational validation of the patient's frustration due to the effort of raising a child can be key to continuing the treatment.

In order to actually answer the patient's question "Can you give me Adderall or Xanax?" the psychiatrist considers how to avoid making the patient go through a loss of power. Since simply not giving Adderall or Xanax may make the patient go through a loss of power, a solution may be to genuinely portray the psychiatrist as being the one without power: "I would give you Adderall or Xanax, but my hands are tied. These are controlled substances and more evidence is required to back up a diagnosis."

Hearing the psychiatrist acknowledge being without power alters the power balance between the patient and the psychiatrist, filling the patient up with a sense of power relative to the powerlessness of the psychiatrist. This may blindside the patient to the patient's own lack of power to get Adderall or Xanax, in turn avoiding treatment disengagement by the patient. At least the patient feels powerful relative to something—here, relative to the powerless psychiatrist.

While the effect of the comments by the psychiatrist on the power dynamic with a patient can be like water putting out a fire, sometimes the effect of the same comments can be like gas on fire, depending on how the comments of the psychiatrist are heard by the patient. No wonder that working with silent patients under anesthesia has its own appeal with psychiatrists.

ECT

Sure enough, there is a form of psychiatric treatment involving just that: silent patients under anesthesia. ECT, or electroconvulsive therapy, is a psychiatric treatment of silent patients under anesthesia—an attempt to electrically enhance the regeneration of brain cells, by passing electrical current through the brain.

It has long been established that ECT works for some patients. But for whom? Of those ECT would work for, some might instead over-rely on medications, or may be too sick to ask for help, or may choose family support instead of psychiatric treatment.

Of the patients that ECT will not work for, some can still fall for having it anyway. This can happen, for instance, when two conditions are met: ECT pays better than other treatment methods, and the patient has a dependent relationship with the psychiatrist. To avoid feeling rejected, the dependent patient might agree to whatever treatment is recommended by the psychiatrist, ECT included. The problem is that the dependent patient will likely feel good even if playing cards with the psychiatrist instead of having ECT.

The dependent patient is inclined to *believe* the method of treatment proposed by the psychiatrist works—in our example, ECT. This way, the patient is more likely to show up at the next session. Wait a minute … *belief?* The issue here is that attention from the psychiatrist is what works, by infusing mental health into the patient-psychiatrist relationship, which is not the same with achieving brain health by way of the ECT. Mental health can mask brain illness. And yet, for mental health alone, attention can be enough, without ECT.

For brain health to be the target, having someone other than the psychiatrist measure whether ECT works tries to take out from the equation the mental health of the patient-psychiatrist relationship. In contrast, when the psychiatrist asks the patient at the follow up session "how do you feel?", the patient gives an answer colored by the patient-psychiatrist relationship. With an intense coloring due to, for instance, a dependent relationship patient-psychiatrist, the answer may reflect the mental coloring more than the brain reality.

And yet, having someone else than the psychiatrist measure whether ECT works tends to cost money.

✳

Grant funds for ECT research are allocated by a system with stringent standards, which are more likely to be met by psychiatrists whose professional lives have been dedicated to ECT—usually by being part of a big organization that invests lots of money in ECT development.

Will these psychiatrists want to come forward with critical results about ECT? Criticizing ECT would jeopardize the method of treatment *believed* in by the psychiatrists enough to build professional lives around it. Wait a minute … *belief* again? And what will the organization investing lots of money in ECT development say when the results are critical of ECT, putting at risk the capital invested?

On the other hand, psychiatrists with less interest in ECT are in a better position to criticize ECT when necessary, but they may not be part of a big organization that invests lots of money in ECT development and, consequently, they may not meet the stringent standards for grant funding. With no money there is no research, so they can't criticize ECT with the credibility that research outcome offers.

RESEARCH

This brings us to how the psychiatrist can manipulate the research outcome in order to validate the treatment that the psychiatrist already believes in. Note that the manipulation is not necessarily intentional, but may be the result of wishful thinking on the part of the psychiatrist.

Usually psychiatric research examines a sample of patients and generalizes the findings from the sample to a large pool of patients. One way to manipulate the research outcome is by sample selection bias. For instance, the psychiatrist might select patients for the sample who believe in the same treatment the psychiatrist already believes in. By portraying the sample as being representative of a large pool of patients, the outcome of research is biased in favor of the treatment believed in by both the psychiatrist and the cherry-picked patients.

In addition to a bias due to sample selection, a hidden factor can turn a sample from being representative into being biased, when three conditions are met: the factor is unknown to the psychiatrist; the factor influences the findings in the sample; and, the factor is not present outside the sample. Such a hidden factor is called a confounding factor. An example is a recent frustrating event undisclosed to the psychiatrist, influencing the findings in the sample, and not present outside the sample.

Separately from a sample selection bias and a confounding factor, the outcome of research can be manipulated when generalizing the findings from the sample to a large pool of patients. Let's look at the example of an ECT study with a sample of ten patients. The researcher observes that ECT administered with a certain frequency works in three patients in the sample and does not work in seven patients in the sample. This is a clear-cut finding within the sample. The question aimed by the study is: does ECT administered with the same frequency work in a large pool of patients?

Without administering ECT to everyone in the large pool, the researcher does not know how the large pool responds to ECT. Because administering ECT to everyone in the large pool is usually not practical, the aim of the study—to know if ECT works in a large pool of patients—cannot be achieved in full. Consequently, the aim of the study is lowered. The aim of certitude having been found too high, the next lower aim is likelihood: not knowing for sure if the sample findings can be replicated in a large pool of patients, the researcher looks at the likelihood of replication, as a method to generalize the sample findings.

The likelihood is a probability, calculated as following: the researcher runs the sample findings—that three out of ten patients responded in the sample—through a sausage machine, a mathematical formula. On one end of the sausage machine the researcher inserts the ingredients, that three out of ten sample patients responded, and on the other end comes out … the sausage, a new numerical result. To determine if long enough, the sausage, the new numerical result, is then compared with a sausage already considered long enough by convention, a threshold of significance.

Thus, the sausage machine, a mathematical formula, calculates the likelihood of replicating the sample findings in a large pool, by a formula-driven probability. When the probability turns out high enough, the likelihood of replicating the sample findings in a large pool is deemed significant.

The issue here is that, to deem significant the likelihood of replicating the sample findings in a large pool depends on two factors: 1. the sausage machine used—the formula; and, 2. the threshold of significance—the sausage considered long-enough by convention, to compare the just-made sausage against. Switching the significance of the calculated probability between "not significant" and "significant" is possible by changing the sausage machine (the formula), or the threshold of significance (against which the calculated probability is compared to), or both.

In other words, the *method*, consisting of the formula and the threshold of significance, to sort out the likelihood of replicating the sample findings in a large pool, influences the very *substance* of the matter: the likelihood itself. By manipulating the method—the formula, the threshold of significance, or both—the researcher can turn on its head the outcome calculated by the method: the significance of the sample findings for a large pool.

Not to mention that the significance is merely statistical. But the public at large does not grasp statistics very well. Blindsided by the word "significance," the public at large tends to ignore statistics, expanding "significance" from being merely relative to statistics (in reality) to being absolute (in the perception of the public at large).

SELF

While statistics speak to a pool of brains, every individual is unique. The uniqueness of every individual comes up in the conversation with the psychiatrist about the self of the individual.

What is the "self"?

Unconscious thoughts are not within self-control but are part of the self; in other words, the self has an out-of-control part: the unconscious thoughts. Psychiatry established a long time ago that unconscious thoughts influence behavior, despite the society expecting behavior to be in self-control. Therefore, a tension comes up between the out-of-control part of the self—the unconscious thoughts—and the behavior, expected by the society to be in self-control despite the influence of the unconscious thoughts.

Conscious values are part of the self too, and they influence behavior as well.

The brain function that negotiates between the unconscious thoughts and the conscious values is called the ego.

The self is made of the unconscious thoughts, the conscious values and the ego.

The unconscious thoughts are impersonal, like "I am hungry" or "I am thirsty." The conscious values are impersonal too, like "Life is good" or "Going to college is good." This reminds me one more time of the game of chess, where each player has the same pieces to play with. The unconscious thoughts and the conscious values are the chess pieces on the table, impersonal to the player.

Playing chess, however, can be done in a practically infinite number of ways, by combining the pieces on the table. It makes how chess is played rather unique, personal to the player. Likewise, the ego—the negotiation function between the unconscious thoughts and the conscious values—becomes rather unique, personal to the individual, because of the practically infinite ways in which the negotiation can be done.

The self is both impersonal, alike between individuals (the unconscious thoughts and the conscious values) and personal, rather unique for the individual (the ego function).

✳

As the self interacts with the environment, how does the environment change the self?

The environment changes the self at two main points of interaction:

1. The unconscious thoughts, coming and going under the influence of the environment; and,
2. The conscious values, under the influence of external values from the environment; e.g. the conscious value "Going to college is good" can be modified by the external value "Freedom from debt is better."

As opposed to the unconscious thoughts and the conscious values, the ego does not interact *directly* with the environment. Instead, the ego remains within the self, not *directly* connected to the environment.

The self is least vulnerable to the environment within the ego, and is most vulnerable within the unconscious thoughts and the conscious values. Through the strength of the ego, the self withstands the interaction with the environment. In a reductionist manner, a person is the ego, at the core. The rest is impersonal layers of unconscious thoughts and conscious values, which come and go with the environment. The ego is personal and negotiates between the impersonal unconscious thoughts and the impersonal conscious values.

✳

What is impersonal to one brain becomes, through the mind formed between two brains, personal to the other brain of the mind. For instance:

1. Bullying behavior originates impersonally, from either an unconscious thought of aggression, or a conscious value of disrespect;
2. The bully is connected through the mind with the victim;
3. Bullying behavior stretches the ego of the victim to the point of rupture, when the ego of the victim struggles to negotiate between the victim's unconscious thought of reactive aggression and the victim's conscious value of friendship. By damaging the ego of the victim, bullying becomes personal to the victim, because the ego is personal to the victim.

On the opposite side:

1. Caregiving behavior originates impersonally, from either an unconscious thought of nurturing, parental-like, or from a conscious value of kindness;
2. The caregiver is connected through the mind with the patient receiving the caregiving;
3. Caregiving behavior sustains the ego of the patient, offering it strength. By strengthening the ego of the patient, supportive caregiving becomes personal to the patient, because the ego is personal to the patient.

Words spoken impersonally can have personal power over the listener. What is impersonal to one brain has personal power, through the mind, over the other brain.

✳

With responsibility attaching to behavior, each person is called to increase self-control. The increase in self-control can be done in two steps:

- step one: raising unconscious thoughts to conscious, through *increased awareness*—performed by the ego; and,
- step two: dealing with the conscious thoughts by the conscious values in a *decision-making process*—performed by the ego (the process of decision-making is not to be confounded with the substantive basis on which a decision is made).

Unfortunately, the ability of the ego to negotiate between unconscious thoughts and conscious values in the two steps (increased awareness and then decision-making process) can be seriously weakened by brain illness. Imagine the ego as a vertical crane, looking into a deep hole where the unconscious thoughts are.

- a sick ego due to brain illness may be long enough to raise unconscious thoughts to conscious (step one ok), but does not know what to do with the now conscious thoughts, due to an impaired decision-making process (step two damaged). This happens in dementia for instance; or,
- a sick ego due to brain illness may be short enough to not be able to raise unconscious thoughts to conscious (step one damaged). This happens in kleptomania for instance. If the unconscious thoughts of stealing in kleptomania would be raised to conscious before being noticed through stealing behavior, they would be dealt with by the conscious values which do not allow stealing in kleptomania (step two would be ok, but can't be achieved due to step one being damaged).

As opposed to in brain illness, the ego in theft is capable of raising unconscious thoughts to conscious (step one ok), and the conscious thoughts of stealing can be negotiated against the conscious values (step two ok)—except the conscious values in theft allow stealing themselves, which results in no need to negotiate (step two not performed). Since the brain is capable to perform both steps one and two, theft is not a brain illness. In theft, thoughts are aligned to values, making the ego useless, despite being capable.

In conclusion, in theft the ego function is intact, while in brain illness the ego function is damaged (for instance, step one is damaged in the brain illness of kleptomania; step two is damaged in the brain illness of dementia).

In general, what keeps the thoughts of stealing under control is the ego function. When the ego function disengages, whether by illness (without self-control) or by alignment of the values to the thoughts of stealing (with self-control), no control is left over the thoughts of stealing. In itself, a thought of stealing is uncontrollable, but for the ego function.

Interestingly, an intact ego function can sophisticate the act of stealing (e.g. premeditation in theft), while a damaged ego function may keep the act of stealing simple (e.g. impulsivity without premeditation in kleptomania).

＊

While an ego not performing step two (decision-making process) hints to a brain illness, an ego not performing step one (of raising unconscious thoughts to conscious) can be unrelated to brain illness. For instance, the unconscious thought of aggression not raising to conscious happens often in the absence of brain illness, like in competitive sports.

To discharge aggression, a raw discharge is not encouraged this day and age. Thus, a masked discharge of aggression becomes necessary.

Competitive sports can be a convenient masked discharge of aggression. Note that the emphasis is on the word "competitive." Hidden behind the competitiveness is the psychological need to make the adversary lose—a form of masked aggression, by taking away what the adversary is losing. The winner takes the trophy, dominates, shows off, while the loser is left with a long face.

The spectator of sports is a competitor by proxy, discharging aggression *unconsciously* toward the loser, under the cover of consciously claiming to enjoy the sport. Make the sport noncompetitive and many people in the audience will likely stop watching. In noncompetitive sports nobody loses, so the unconscious aggression is not relieved.

Being good at something is not enough for the unconscious. Being "better than another" is what the unconscious is after, to discharge aggression *toward* the adversary. The real deal for the unconscious is relationship-based: an adversary is necessary to successfully discharge the aggression. That's when the unconscious feels good, fulfilled, freed.

In general, the masking of aggression is accomplished by interposing, between the self and the adversary, a set of rules stemming from shared conscious values that allow the discharge of unconscious aggression in a controlled manner from the self to the adversary. Examples of such a set of rules stemming from shared conscious values are the rules of: a competitive sport, a competitive beauty pageant, a scientific competition, or an artistic competition.

❋

In an attempt to fit the unconscious aggression within the set of rules stemming from the shared conscious values, the ego can stretch too thin. When the ego stretches too thin, the unconscious aggression can overcome the set of rules. Consequently, the aggression erupts outside the shared conscious values, like when a competitor punches the adversary in the face.

Then a psychiatrist is called to evaluate whether brain illness is at play. No wonder psychiatrists have tried over time to map where the aggression is within the brain.

IGUANA

Science describes three layers of the brain: the bottom brain (the instinctual brain), the middle brain (the emotional brain), and the top brain (the abstract brain, where the cognitive rules are).

For the sake of simplicity, let's draw a horizontal line separating the top brain from the middle brain. This leaves the top brain on top—where the cognitive rules are—and a brain underneath, made of the middle brain and the bottom brain. The brain underneath is an advanced form of instinctual brain, now equipped with emotions. Let's call it the iguana brain.

The top brain and the iguana brain are like two boxers fighting in a ring. A dynamic tension occurs between the top brain and the iguana brain. For instance, getting seriously frustrated with another person generates, by lighting up neurons in the iguana brain, an impulse to throw a punch. On the other hand, the top brain hits the brake on the impulse to throw a punch, because throwing a punch is in conflict with the cognitive rules of consequences in the top brain. The cognitive rules of the top brain operate like computer software, filtering the raw impulses from the iguana brain.

✳

Because the iguana brain includes the emotional brain in addition to the instinctual brain, the suffering is located in the iguana brain, being an emotion.

An illness of the brain, through symptoms and dysfunction, brings the emotion of suffering to the place where suffering occurs: in the iguana brain. This is independent of where the illness is located in the brain. For instance, schizophrenia brings suffering to the iguana brain, independent of the location of hallucinations and delusions outside of the iguana brain.

Since suffering is a risk factor for worsening a psychiatric illness, the job of the psychiatrist includes to alleviate suffering, which requires sometimes the psychiatrist to root for the patient's iguana brain. Below are three examples of when the psychiatrist roots for the iguana brain of the patient.

*

The first example involves disability benefits. A patient comes to the psychiatrist's office, unsure whether to talk about suffering from not being able to work, or simply to continue looking for work. There is a tension between the top brain, *pushing* the patient to "try harder" to find work, and the iguana brain, *suffering* from lack of work. When unresolved, the tension between the top brain and the iguana brain makes the brain run out of chemical transmitters between neurons, exhausting the brain, and leading to depression.

To mitigate depression, the patient applies for disability benefits. However, from the disability judge's perspective, a determination needs to be made on whether the patient meets the legal definition of disability. In that, the disability judge may ask the psychiatrist for feedback. But the allegiance of the psychiatrist is to the patient, not to the disability judge.

The judge has to answer the question "is the patient legally disabled?", while the psychiatrist has to merely answer the question "does the patient have an *apparent* disability due to a psychiatric illness?".

Note that a loop forms between brain illness, the financial damage due to the dysfunction in brain illness, the subsequent suffering, and the risk of worsening of brain illness by way of the suffering. The psychiatrist is motivated to break the loop by supporting a disability benefits application, which in turns improves the finances, making the suffering less, and thus reducing the risk of worsening the brain illness as a result of suffering.

By decreasing the suffering through supporting a disability benefits application, the psychiatrist in essence roots for the iguana brain, where the emotion of suffering occurs. The psychiatrist becomes an advocate for the patient, looking at presenting to the court the clinical facts in the light most favorable to the patient, short of lying.

But why not present the clinical facts "objectively"? Well, because to objectively present the subjective matters of psychiatry is science fiction. These matters are *subjective*. As said before, the *subjectivity* of the patient is the *objectivity* the psychiatrist has to work with. The psychiatrist does not work within what society calls "objective reality"—a logical, rational reality, widely accepted. Instead, the psychiatrist works within a person's skewed subjective reality, colored emotionally, sometimes rather unique to the patient—the reality in which the patient operates.

Even with a psychiatrist supporting the disability application, the judge might still deny it. For the judge, disability is essentially a term of art—the term has a legal meaning not identical with the common use of the word among people that do not have legal training. Disability for the judge is defined by the underpinnings building up to the threshold of being *legally* disabled—like the letter of the statute, the spirit of the statute, the case precedents, and the rules and regulations of administrative agencies.

That is not the same for the psychiatrist, who cannot be expected to apply the law to facts, even clinical facts, given that applying the law is in general a process that requires legal training—how else can someone appreciate, for instance, the weight of a legal case precedent not fitting perfectly within the facts at hand?

✳

The second example of when the psychiatrist roots for the patient's iguana brain follows here. Consider a patient whose iguana brain suffers because of an illness that makes the patient want to live alone in the house, even though immediate family members live there too. Asking the immediate family members to move out seems like a relief to the patient.

Even though the psychiatrist may logically reason that the family staying in the house is good for the patient, the allegiance of the psychiatrist is to the patient, not to the family. The psychiatrist has to decide whether to encourage a gesture hurting the family cohesion for the sake of healing: the patient essentially kicking family out of the house. When brain illness demands that the patient lives alone, the burden from brain illness is lessened once the family is out of the house. Then, the patient gets a step closer to healing.

The psychiatrist might struggle with the morality of the patient's choice. But a judgmental psychiatrist can hurt the patient's opportunity to get better. When getting better, the patient goes from the non-autonomous phase driven automatically by the illness and not in self-control, to the autonomous phase driven by choice and in self-control. Then, the patient is better able to guide the behavior by values, instead of being derailed by the illness.

The psychiatrist can facilitate the healing of the patient by supporting the selfish iguana brain of the patient, through supporting the questionable behavior of asking the family to leave. With suffering being a risk factor in worsening the psychiatric illness, the decrease in suffering by the patient living alone facilitates healing independent of how other family members feel.

＊

The third example of when the psychiatrist roots for the patient's iguana brain is when the psychiatrist works in prison. The top brain makes the incarcerated patient repent. The iguana brain makes the incarcerated patient feel comfortable. The tension between repenting and feeling comfortable is the tension between the top brain and the iguana brain.

When the tension is too high, the brain gets exhausted, running out of chemical transmitters between neurons. Then, depression sinks in. The depression combined with the repentance from the top brain stop the iguana brain from feeling comfortable. Instead, the iguana brain begins to feel bad. This can make the depression worse. A loop forms between the worsening of the depression and the feeling bad of the iguana brain.

To break the loop, the job of the psychiatrist includes helping the incarcerated patient push back for the moment on the repentance from the top brain, as repentance is a factor in making the iguana brain feel bad, and therefore in worsening the depression. The psychiatrist wants the iguana brain to feel comfortable again, as a step toward healing.

In general, suffering from being incarcerated is meant by the justice system to impact the substantive decision-making of the prisoner, autonomous and within self-control, but not to eliminate the process of decision-making by inducing depression, non-autonomous and not within self-control. There is a limit to suffering: the justice system hires the prison psychiatrist to treat the inmate that gets depressed.

After the treatment leads the inmate to more autonomy and self-control, the top brain is in a better position to change the substantive decision-making of the inmate, without losing the process of decision-making.

✳

Is the iguana brain egoistic? Very much so.

The iguana brain must satisfy two separate kind of demands: the primitive instinctual demands—egoistic—and the emotional demands—seeking companionship. Without satisfying the primitive instinctual demands, the iguana brain suffers, despite companionship. To not suffer, the iguana brain requires, in addition to companionship, the satisfaction of the primitive instinctual demands. An example is a marriage lived in full, much better for the *iguana* brain than a platonic marriage.

✳

The iguana brain is not at fault for lighting up thousands of neurons at the sight of a physically beautiful person, nor for remaining relatively dark at the sight of a person physically not so beautiful. While the physical beauty of a person may be relative to perception, the iguana brain tends to regard symmetrical facial features as beautiful, as opposed to asymmetrical facial features. Take a look at some of the top earners in Hollywood, where physical beauty matters. What excites the iguana brain is not fair play. Fairness comes from above.

＊

The iguana brain is guided by the cognitive rules of the top brain, which allows the individual to share the environment peacefully with others.

The guidance of the iguana brain by the top brain can be fractured by brain illness. Automatic and non-autonomous, brain illness can take over the guidance of behavior from the top brain. This puts the patient at risk for new behaviors, usually not present but for brain illness. Paul Tournier calls them in his book, *The Meaning of Persons*, the automatisms driven by the underlying brain illness, distinguished from intentional choices of behavior.

Examples of new behaviors usually not present but for brain illness include: barricading oneself in a room when paranoid; not getting out of bed when depressed; engaging in a shopping spree when manic; avoiding public places when anxious; or, urinating in public when drinking excessive amounts of alcohol.

Such new behaviors during brain illness bring comfort to the iguana brain, by freeing the brain to manifest itself by virtue of the brain illness, stressing instead the minds the individual is part of.

＊

In general, what keeps the iguana brain comfortable does not necessarily overlap with what keeps the surrounding minds alive. Making the iguana brain too comfortable by doing what the iguana brain wants can alienate the surrounding minds, ultimately killing them. Then, the iguana brain suffers the loss of the surrounding minds, suffering being and emotion, and thus happening in the iguana brain. A chain of effects takes place, in which the keeping of the iguana brain too comfortable destroys minds, followed by isolation, followed by suffering of the iguana brain in isolation, which can ultimately induce brain illness as a result of keeping the iguana brain too comfortable.

Therefore, the iguana brain cannot afford to simply ignore the surrounding minds. Instead, the iguana brain uses the top brain, where the cognitive rules are, to attempt to find a way of getting what it wants and at the same time to avoid killing the surrounding minds. The top brain is in the position to overrule the iguana brain when necessary to keep the surrounding minds healthy. This in turn makes the iguana brain go through suffering.

The suffering by the iguana brain can be bearable, or unbearable. Take the example of a personal sacrifice on behalf of a relationship, like the sleepless nights of a mother for a child, or, if you wish, the effort of an artist to bring a magnum opus to the audience. The choice of the personal sacrifice by the top brain keeps the mind alive between the parent and the child, respectively between the artist and the audience. When the suffering of the iguana brain is bearable, the overruling of the iguana brain by the top brain works. When the suffering of the iguana brain is unbearable, the overruling of the iguana brain by the top brain invites brain illness to the table.

Once brain illness occurs as a result of an unbearable overruling of the iguana brain by the top brain, brain illness is prone to put a strain on the minds surrounding the brain. A chain of effects takes place, in which brain illness destroys the surrounding minds, followed by isolation, followed by suffering of the iguana brain in isolation, which increases the risk of worsening the brain illness already present as a result of an unbearable overruling of the iguana brain by the top brain.

To reduce the likelihood of brain illness due to the iguana brain being too comfortable or due to the iguana brain being unbearably overruled by the top brain, the top brain needs to refine the power dynamic with the iguana brain, in such a way that the iguana brain is not too comfortable, but is also not unbearably overruled. The top brain then needs to engage in a balancing act with the iguana brain, to follow the cognitive rules while stopping short of facilitating brain illness.

VALUES

The cognitive rules of the top brain stem from conscious values. When the values change, the cognitive rules change. For instance, with a change in values, a previous cognitive rule of "this is unacceptable" can change to a new cognitive rule of "this is acceptable". Consequently, changing values becomes tempting in order to decrease the tension between the cognitive rules of the top brain and the iguana brain.

Fortunately, values don't have to change to opposite values. Sometimes, values can change to similar but more affordable values. Example: when the sleepless nights of the mother for the child open the door to brain illness for the mother, the value of taking care of the child does not have to change to the opposite value of letting the child uncared for. Instead, the value of taking care of the child can change to a similar, but more affordable value: wake up the father. This gets the father involved in caring for the child at night, while the mother can have a well-deserved rest.

Not only people, but systems too, like medicine and psychiatry, operate with values.

Medicine operates in an environment of values, being more than a technique. For instance, prolonging life in an irreversible permanent vegetative state is a matter of values: length of life versus quality of life, determining how far to extend life support. Placing an implant in a healthy butt is a matter of values: form versus medical need, determining what is a medical act. Prescribing narcotics for a non-cancerous chronic pain is a matter of values: a pain-free state versus avoidance of the risk of medication dependence, determining how far narcotics are used.

Psychiatry operates in an environment of values too. For instance, not telling others what the patient says is a matter of values: confidentiality versus safety, determining the information that can still be disclosed. Continuing a session with a belligerent patient is a matter of values: compassion versus accountability, determining how far the belligerence of the patient can go. Following a patient's request for a particular medication is a matter of values: accuracy versus customer satisfaction, determining how far psychiatry can accommodate a patient's request.

Despite the appearance of being a science with inner consistency, psychiatry is not self-contained enough to be isolated from an environment of values. Being under the influence of an environment of values, psychiatry cannot escape being shaped by values.

A common complaint from patients not depressed anymore as a result of antidepressant treatment is that, even though the depression is now gone, a sense of aloofness sinks in. With a blank stare, the formerly depressed patients wonder what next. With no answer in sight, they turn to the psychiatrist for guidance beyond the depression already gone. Given that psychiatry cannot escape being shaped by values, the psychiatrist owes to the patient to discuss values.

✸

The psychiatrist needs to be careful when regarding psychiatric matters through values inherent to the psychiatrist. For instance, when a patient complains of being angry with a neighbor, a Christian psychiatrist may be tempted to say "forgive your neighbor". That would be a nice attempt to remind the patient of the value of forgiveness by way of the psychiatrist-patient mind. However, when self-control over behavior is lacking due to brain illness, the choice of the patient to steer behavior toward the value of forgiveness is gone. The Christian thing to do then is to treat the patient psychiatrically, for instance by starting a mood stabilizer, instead of merely giving an advice that the patient simply cannot follow.

Note that, in reality, self-control over behavior is not black and white, either present or absent. Instead, self-control over behavior can be partial. Assessing how much self-control the patient has over behavior is important in order to prioritize treatment versus advice. While putting a diagnosis is well emphasized in medical training, assessing the degree of self-control of the patient over behavior is underemphasized, despite being important in the evaluation of dangerousness in general.

＊

Evaluating dangerousness is a risk assessment, but not a guaranteed prediction. The future cannot be predicted by psychiatry. This brings us to nothing less than quantum mechanics. The model of parallel histories from quantum mechanics can be useful to psychiatry: with uncertainty at play, attempting to compress all parallel realities into one reality is not possible in a given moment in time. Then, parallel realities become potential realities. With a future not crystal clear, the passing of time will tell which parallel reality materializes.

＊

Psychiatry evolved historically as a tool to address illness, not as a narrative to the truth. And yet, to pay attention to the mind—not only to the brain—psychiatry needs to raise from the condition of being a tool, to mindful thinking. As the psychiatrist is not supposed to lie, the mindful thinking becomes a narrative going to where the psychiatrist senses the truth is. When the truth is unknown, a value becomes the compass to the truth.

Different values, some opposite to each other, populate psychiatry nowadays. Thus, mindful thinking in psychiatry is constrained by the tension between values in psychiatry. To release the tension, value-driven slices of psychiatry come into play. Let's call them parallel psychiatries. Each parallel psychiatry determines where the truth is by the compass to the truth—the value driving that parallel psychiatry.

Note that psychiatry does not have the doctrinal force to explain in perspective competing values, like a religion would. Instead, psychiatry is a science limited to accept a division between parallel realities that the science observes. In that division between parallel realities, the individual psychiatrist can assume within a value-driven parallel psychiatry the otherwise unknown truth.

When feeling good is elevated to be a value, the truth comes in the form of doing what the person feels like doing. But letting feelings alone be in control of behavior can mean not being in control, as feelings can be driven from the unconscious, where control is lost. Reason, if applied, can disconnect behavior from feelings.

When doing the right thing is elevated to be a value, the truth comes in the form of doing the right thing. But doing the right thing while ignoring the unconscious can feel wrong. Feelings, if applied, can disconnect the behavior from doing the right thing.

When self-empowerment is elevated to be a value, the truth comes in the form of the betterment of self. But the capacity of some brains for betterment of self is lost due to brain illness. To empower a brain without capacity for betterment of self, a separate source of power than the self is necessary: the mind.

When religion is elevated to be a value, the truth comes in the form of doing what the religion expects. But doing what the religion expects can enter in conflict with the secular society, which has different behavioral expectations than the religion has.

✳

As patients ask for guidance in solving moral dilemmas, the respect for psychiatry is at stake. For instance, when a patient is slapped by someone, what does psychiatry encourage the patient to do?

The answer depends on which parallel psychiatry we are in. For instance, a parallel psychiatry driven by the value of feeling good can say to the patient to do what the patient feels like doing. Or, a parallel psychiatry driven by the value of doing the right thing can say to press charges against the assailant. Or, a parallel psychiatry driven by the value of self-empowerment can say to slap back in self-defense. Or, a parallel psychiatry driven by the value of religion can say to, for instance, turn the other cheek.

By listening to the value of the parallel psychiatry, the self puts the ego function in the position of negotiating between the unconscious thought of aggression after being slapped and the conscious value driving the parallel psychiatry.

Note that the ego function does not operate in a vacuum. Instead, the ego function operates in a medium of feelings. The unconscious thoughts and the conscious values are surrounded by feelings, like bones being surrounded by muscles. For example, the feeling surrounding the unconscious aggression after being slapped is usually anger. The ego function can act consistent with anger, like when letting the slapped individual either do what feels like doing, or pressing charges, or slap back in self-defense. Also, the ego function can act opposite to the anger, like when making the slapped individual turn the other cheek.

When turning the other cheek despite an antagonistic emotion, the ego function operates on behalf of the mind, not on behalf of the brain anymore—it makes the jump from being part of the old self to being part of the mind instead. The new self is now centered on the mind, not on the brain, despite the antagonistic emotion being a drag. Turning the other cheek slides the behavior from retaliatory for the brain to rehabilitative for the mind.

✹

Hopefully no parallel psychiatry will tell a patient slapped on the cheek to shoot the slapper. And yet, parts of the world allow just about everybody to carry a gun, including patients. This raises the moral dilemma of what patients can do with guns.

Some people say that guns can stop misbehavior. But guns can also be a way for patients to express frustration. Statistics indicate that, if everyone has access to guns, a number of distressed people almost certainly will be violently harmful. The math of the random biological propensity to violence is clear, as the news channels show from time to time.

A judge once said that brain illness and evil are not mutually exclusive. Unfortunately, the judge was right: the mere presence of brain illness does not exclude evil. Moreover, by facilitating a loss of power to reject temptation, brain illness can make a person vulnerable to listening to the evil.

The violent impact of loss of self-control can be amplified by guns. Blindly allowing anyone the right to bear arms disenfranchises those who lose self-control of behavior as a result of brain illness. These people can fire the guns *uncontrollably*.

A strong society is as vulnerable as its weak spots. When the strength of the law granting the right to bear arms becomes a weak spot of the society, by allowing in effect an uncontrollable firing of weapons when control over weapons is lost to brain illness, then a line ought to be drawn for who can bear a weapon. It's not a matter of taking a right away. It's a matter of acknowledging the difference between the law of the land and the law of nature. For people with brain illness that lack control over behavior, the right to bear arms is essentially the right to fire weapons uncontrollably.

It's like saying that everybody has the right to drive a car without needing a driver's license, just because a long time ago everybody had the right to ride a horse. Some people are simply not capable of driving a car without getting into an accident. Perhaps it's a matter of enough accidents, to reach the tipping point of drawing a line.

Where to draw the line depends on who cannot control the guns. But who cannot control the guns is far from static, *changing in time* under the influence of factors that include: the state of the brain, the state of the mind, and the circumstances.

For comparison, recall the subject matter in engineering called Strength of Materials, dealing with the ability of any material, for instance steel, to withstand a pressure put on the material without snapping, like a weight. To determine the resistance, the material is tested under a pressure. Just because a material does not snap today, in the absence of a pressure, does not mean the material will not snap tomorrow, under a pressure.

In this light, the cutoff line to efficiently prevent an uncontrolled use of guns is at *more* than who cannot control guns *today*; the cutoff line includes those who will lose control tomorrow, under a pressure, despite having control today, under no pressure. Note that with a high enough pressure, a lot of people may lose control, making it impossible to determine who will snap tomorrow, under what exact pressure, despite good control today under no pressure. The uncertainty in the prediction leads to snapping tomorrow by practically random people, who have good control today. This pushes the line of who can bear arms *without* snapping into uncontrollable firing to organized groups with intrinsic supervisory mechanisms, like an army.

✳

Are values necessary in psychiatry? Patients often ask, "Doc, what do you advise I believe in?" Traditionally, the psychiatrist does not advise patients what to believe in. So, wouldn't it be simpler to have a valueless psychiatry?

The question is a tricky one because values come to the brain by way of the mind, and therefore must be dealt with by the brain independent of whether they are necessary. The better question is how to deal with the encountered values.

For instance, bad values come to the patient by way of the mind. Let's call it the bad mind. What can stop the bad mind from influencing the brain of the patient? A good functioning of the patient's brain can. But when brain illness kicks in, the brain can lose enough functioning for the bad mind to win over the brain. Then, to push back against the bad mind, more is needed than a dysfunctional brain secondary to illness. A separate good mind is then necessary.

By way of the mind, a value can define the trajectory of the ill brain that struggles to retain control over the trajectory. Competition between values ensues. On the other hand, accepting multiple values without competition can blur the differences between values, diluting a value's ability to define the trajectory of the ill brain. In an extreme, when every value is neutralized by an opposite value of the same weight, a value's ability to define the trajectory of the ill brain is lost, and consequently the ill brain becomes unhinged.

When a value has weight by way of the mind, the value gives the ill brain a sense of meaning within the mind. To maintain the sense of meaning, a value-based psychiatry becomes inescapable. The answer from the psychiatrist to the patient on the question "Doc, what do you advise I believe in?" amounts to a value with the weight of certitude by way of the mind.

✳

Beyond giving a sense of meaning within the mind, a value is necessary for the brain to keep in check unconscious thoughts. The unconscious fear of death is kept in check by, for instance, the value of healthiness. The unconscious fear of personal rejection is kept in check by, for instance, the value of doing a good job—resulting in being accepted instead of being rejected. The unconscious aggression is kept in check by, for instance, becoming a surgeon—a good way to put knives in people. The unconscious anger is kept in check by, for instance, the value of being a good debater.

Values are necessary to maintain a sense of sanity. They keep in check unconscious thoughts that threaten to take over the self. Without values, unconscious thoughts raise to consciousness unopposed, taking over what is left of the self. Then, the unhinged unconscious thoughts, now conscious, scare the patient, who foresees what will happen if acting on them. The effect on the patient is devastating, with a severe increase in distress by way of, for instance, panic attacks, paranoid thinking, suicidal thoughts, homicidal thoughts, obsessive thoughts, or psychotic melancholia. Ultimately, the patient can lose control over the behavior.

When having to walk through a dark cave, the ego function tries to determine what lies ahead by taking steps, like logic, the bringing of surroundings into awareness, and weighing of feelings. If not able to see ahead, the ego function has one more step to use: the assumption of a value that says the dark cave is safe despite not knowing for sure. Death is an example of a dark cave where, against all odds, the assumption of a value can compensate for the fear of what lies ahead.

Perhaps referring to the importance of values in keeping unconscious thoughts in check, St. Ignatius once said, "If the guide God chooses for you to follow were only a little dog, you should not complain, but at God's command follow it willingly and gladly".

REWARDS

Life is not made of dark caves all day long, however. Life can be mostly a succession of plain fields, with mountains and valleys here and there. In the geography of life, some destinations are important because they put in action the dopamine pump in the pleasure circuit of the brain—they make the individual feel good. We call these destinations rewards.

Periodic dopamine release through reward attainment is known to be a protective factor against developing brain illness. On the other hand, failing to arrive at destination does not release much dopamine. Consequently, failing to arrive at destination puts the brain at risk of becoming ill, by eliminating the protective factor against developing brain illness that enough dopamine discharge at reward attainment is.

When a person wants to get from point A to point B, two choices are in place: 1) follow the laws of physics; or, 2) fail. It's as simple as that. To master the physical universe, one has to follow the laws of physics. But how important is following the laws of physics in the inner life, where the laws of physics apparently do not apply?

When the laws of physics dictate the trajectory to a reward, the same self-discipline required to follow the laws of physics is needed to attain the reward. When unable to operate within the laws of physics, rewards are missed, and consequently not much dopamine discharge takes place. Thus, failing to follow the laws of physics results in missing the protective factor against brain illness that enough dopamine discharge at reward attainment is. Consequently, failing to follow the laws of physics puts the brain at risk of becoming ill when compared to a brain that follows the laws of physics to successfully attain rewards.

❋

Rewards are stimuli from the environment. The brain is attracted to rewards by anticipating the release of dopamine at reward attainment in the pleasure circuit. The brain is directly attracted by intrinsic rewards—examples include food, water, an intimate contact with a partner. The brain learns to be attracted to extrinsic rewards by association with intrinsic rewards. An example of an extrinsic reward is money, paired by social convention with an intrinsic reward (e.g. "a dollar for a cookie").

Society gives access to rewards to competitive people. What makes an individual competitive is a functional brain. On the other hand, a dysfunctional brain makes an individual less competitive. Some rewards are then out of reach.

Animal observations showed that reaching rewards motivates subjects to search for more rewards. Animal observations also showed that repeated failed attempts to reach rewards leads to learned helplessness. Subjects in these cases later do not pursue rewards within reach should they try.

Similarly, patients can fall into learned helplessness when not competitive enough to follow societal rules in searching for rewards.

Or, patients can find shortcuts to rewards by breaking societal rules. Note that, whether rewards are allowed by the societal rules or not, brains are attracted to rewards naturally. A warm place during snow has a healthy effect on the brain of a homeless person with brain illness, no matter how the homeless individual got to the warm place— whether by following societal rules, or by trespassing.

In general, when an iguana brain hurts from brain illness and a reward has a healthy effect on the iguana brain, asking the patient to give up on the reward in order to follow societal rules may prove too much to ask for. The health that the reward brings to the iguana brain does not even have to be a matter of life and death for the individual to break societal rules; sometimes it happens for much less.

Take a person with bipolar, in an agitated state, with no job. How is this person supposed to find entertainment—a novelty-seeking reward against the discomfort of a boredom amplified by the bipolar? When there's no other way, sneaking into a movie theater gets it done, even though theft of entertainment services is against the societal rules. This is not to condone unlawful behavior. This is to describe what happens in real life from time to time.

*

A reward can replace a mind.

A reward can cut off the mind between two brains, turning each brain to the reward instead of the other brain. Left without the mind, the brain driven by a reward is deprived of the power of the mind. An example includes a business deal between two friends that goes well until they sacrifice the friendship to fight over who will take over the entire business. Here the money as reward replaces the mind once sustaining the friendship, as the money turns from being a means at the disposal of the mind to being a purpose in itself.

*

Business relationships, otherwise important in society, are based on the consideration of money. On the other hand, brain illness leads to a well-established downward economic drift, because brain illness usually comes with dysfunction by definition. To integrate people with brain illness into society requires a reconceptualization of the social contract—how people deal with each other—on nonfinancial terms. It becomes less about how much money can be made off a person, and more about how a person is directly valuable to another person.

Integrating people with brain illness into society is in tension with making money. The resolution comes from prioritizing: a group that prioritizes forming minds can become a mindful community, striving to integrate people with brain illness; a group that prioritizes making money is at risk of turning into a mindless pack of reward producers, left without a need to integrate people with brain illness.

✳

An example subtler than money of a reward cutting off the mind between two brains is a relationship driven by being intimate—a reward—and no more. This is the case of the couple that "resolves all differences in bed". When the same two people stop sharing a bed, the differences between them can crumble the reward-based relationship, absent a mind to resolve the differences.

✳

Another example of a reward that cuts off the mind between two brains is a powerful street drug. The drug chemically releases the dopamine in the pleasure circuit of the brain. As noted already, the drug hijacks the pleasure circuit of the brain, eliminating the need of the brain to function in the surrounding environment in order to experience pleasure.

The mind can be lost to the powerful drug, being attacked from two sides: the brain that uses the powerful drug can lose interest in the mind (as dopamine release is done by the drug, without needing the mind), and the other brain that would be part of the mind can lose interest in the mind (as dopamine is not released in interacting with an erratic brain on the drug).

The brain using the drug may not respond to sweet-talk into giving up the street drug, because the pleasure from the street drug is more intense than the emotion generated by the sweet talk. Sometimes the brain cannot even be reasoned with to give up the street drug, when the brain miscalculates the risk of the consequences of using the street drug. For instance, when brain illness is present, the calculation can be blindsided by four factors:

1. the instant gratification from the street drug (amplified by the frustration over rewards in general not being easily accessible when having brain illness);
2. the dampening of the cognitive abilities due to brain illness;
3. the cognitive side effects of the drug itself;
4. the cognitive side effects of the psychiatric medication treating the brain illness.

Note that even without a previous brain illness, the drug itself can lead to brain illness, for instance by activating dormant genes for brain illness that would not have turned active but for the drug use.

To consider engaging in recovery, sometimes an individual has to go beyond listening to a sweet-talk about giving up the drug, and beyond calculating the risk of consequences of using the drug. Hitting rock bottom is sometimes necessary. The actual life consequences may open the eyes of the individual, by proving without a doubt that using the drug is in fact not worth.

To prevent the patient from hitting rock bottom, an enhanced consequence can be used. An example of an enhanced consequence is the loss of freedom by a court-ordered inpatient rehabilitation treatment. The loss of freedom as an intervention against the use of the drug may tip the balance between the use of and the consequences of using, pushing the patient toward not using. Note that nowadays the loss of freedom as an intervention against using a drug is limited by the societal preference for treatment at will. This is not to advocate for loss of freedom. It is to simply describe the notion of enhanced consequence before hitting rock bottom.

*

We just talked about rewards. Previously, we talked about values. What is the relation between rewards and values? They both seem to be attractors of the brain. Let's delve a bit in how rewards and values compare.

A tension exists between rewards and values. For instance, someone else's food—a reward—is in tension with the value of "do not steal". The reward and the value compete with each other to drive the behavior. The solution to get the reward and at the same time to preserve the value is to follow, when getting the reward, the rules stemming from the value.

But when for a person with brain illness the rules are impossible to follow, then a choice has to be made: miss the reward by sticking with the rules, or get the reward by breaking the rules.

Food, being a necessary reward, presents a challenge when trying to uphold a value over it. Following only the rules stemming from the value consumes energy, which in turn requires food to get the energy from. The ill brain gets hungry after a while, when not attaining food through the rules stemming from the value. Consequently, the rules stemming from the value are now threatened by the laws of nature, which demand the brain to pursue food independent of the value.

This brings up the issue of the discrepancy between the rules stemming from the value and the laws of nature. When "clever" legislators want to pass rules against nature, the question becomes: Who is going to win, the "clever" legislators, or nature?

As noted before, values come to an individual by way of the mind—education in school is a classic example; or, the laws of the land. On the other hand, rewards touch an individual by way of the brain—for instance, food results in a discharge of dopamine in the brain. The mind is in tension with the brain when, at the same time, the mind is attracted by a value in one direction (e.g. "do not steal") and the brain is attracted by a reward in a different direction (e.g. food).

Without values, we are in the land of reward psychiatry, where humans turn into animals. Note that animals do not have the notions of robbery, assault, rape or killing; all come naturally, by way of the Darwinian survival of the fittest. Values, on the other hand, hold in check, within the human, the would-be animal.

Trying to resolve the tension between rewards and values puts values at risk of being switched from values in tension with the rewards, to values in sync with the rewards. For instance, the value "do not steal" being in tension with the reward of food, the alternative value "it's ok to steal" is in sync with the reward of food. Moreover, once a person replaces the value "do not steal" with "it's ok to steal", the lack of tension between the new value and the reward, and the consequent freedom from value to access the reward, reinforces the new value ("it's ok to steal"). This in turn reinforces the behavior of stealing, now congruent with both the new value and the reward.

Brain illness has the potential to lead to moral poverty when facilitating a switch in values in order to change the rules, stemming from values, to attain rewards. The attaining of rewards reinforces the new values, and reinforces the behavior that would have been prohibited by the old values. This behavior can take a life of its own, continuing even after the rewards becomes later accessible under the old values (e.g. after regaining functioning by successful treatment of brain illness).

Moral poverty with its behavior can survive the recovery from brain illness.

＊

When the successful treatment of brain illness leaves the moral poverty of switching values unresolved, moral recovery requires an approach that the psychiatric treatment is simply not traditionally set to achieve.

Examples of approaches to achieve moral recovery include: 1. prison time, to learn by association that the reward gained by breaking rules, such as by stealing, is not worth the stay in prison; 2. a faith-based intervention, where growth within the faith-based mind targets the formation of new values brought along by way of the mind.

Note that both the prison and the faith-based intervention do not address what led to the switch in values in the first place: the dysfunction due to brain illness— an obstacle to finding rewards. Addressing the dysfunction by treating the brain illness is a stepping-stone to finding rewards without having to switch values.

Nonetheless, when brain illness cannot go away through treatment and stays in the way of finding rewards, attaining rewards can still be accomplished without the patient switching values, when the society allows easier access to rewards. For one, making rewards more accessible offers an alternative path to rewards for people with brain illness than switching values. This fosters the integration of people with brain illness into society. Secondly, making rewards more accessible can put a stop to the added suffering of people with brain illness due to not reaching rewards, breaking the loop of brain illness—no rewards—worsening of brain illness through suffering.

＊

Values do not have to be subservient to rewards. The quality of life of a person with brain illness remains an important value in society, despite the cost in rewards to society. As psychiatric treatment tends to improve the ability of the ill brain to form a mind, rewards supporting psychiatric treatment become a means to form a mind, not a purpose replacing a mind.

✳

As a reminder, within the self, the ego function negotiates between the unconscious thoughts and the conscious values.

The environment does not interact directly with the ego function. Therefore, a reward, being a stimulus from the environment, does not interact directly with the ego function.

The environment can interact directly with the unconscious thoughts. Therefore, a reward, being a stimulus from the environment, can interact directly with the unconscious thoughts. This happens, for instance, when the reward generates the unconscious thought "I want this reward", in anticipation of the dopamine pumping in the pleasure circuit of the brain at reward attainment.

The environment can interact directly with the conscious values. Therefore, a reward, being a stimulus from the environment, can interact directly with the conscious values. This happens, for instance, when the reward leverages the alignment of conscious values to the reward, in order to make the pursuit of reward by the self more efficient. At extreme, the reward can go as far as pushing aside the conscious values for the purpose of attaining the reward at all price. Then, the brain gets out of touch with the values for the sake of the reward: the brain is hijacked by the reward.

The hijacking of the brain by the reward risks snapping whatever minds the brain is part of, as the brain does not follow rules stemming from values anymore, ultimately making the brain not only valueless, but mindless too.

✳

Opposite to rewards are counter-rewards, or aversive stimuli, acting on the aversive circuit of the brain, and thus generating avoidance by the brain. They also come in intrinsic and extrinsic forms. The brain directly avoids the intrinsic aversive stimuli—for instance, a very cold air, or a loud noise, or a bitter flavor. The brain learns to avoid extrinsic aversive stimuli by association with intrinsic aversive stimuli, for instance that pushing a button (extrinsic aversive stimulus) makes a loud noise (intrinsic aversive stimulus).

An aversive stimulus leads to avoidant behavior. The dysfunction of brain illness, however, could prevent the brain from sensing the aversive nature of the stimulus; in essence, people with brain illness may disregard the aversive nature of a stimulus, by misperception or miscalculation, in the shadow of the dysfunction of brain illness. Then, a barrage of aversive stimuli tends to follow—for instance, ignoring the aversive nature of breaking the law can be followed by the aversive nature of loss of income, loss of housing, loss of marriage, loss of personal property, and loss of freedom.

But all hope is not lost. When brain illness makes a brain go "blind", by preventing it from recognizing the aversive nature of the stimulus, a mind can still make the aversive stimulus visible to the otherwise "blind" brain.

FRAME

There are limits put in place to what the patient-psychiatrist mind can accomplish. To protect the patient and the integrity of the profession, boundaries are set.

Nonetheless, when suffering in isolation, a lonely patient naturally looks to whoever happens to be around, such as the psychiatrist, for companionship. After all, what could be better to ward off loneliness than, let's say, a person to go to dinner with, maybe watch a movie together, or develop a friendship with?

Boundaries limit the relationship with the psychiatrist, and are in tension with the longing by the patient for a fully developed relationship.

＊

Some patients, if not all, attempt to break boundaries. This may be due to psychological mechanics—such as the projection of feelings from past relations on the psychiatrist—or, due to the longing by patient for a fully developed relationship. Despite attempts to break boundaries by the patient, the relationship with the psychiatrist usually remains in place.

Even though, because of the boundaries, the relationship with a psychiatrist is a limited deal compared to, let's say, a good friendship, it is a deal nonetheless, which is better than no deal at all. In fact, the relationship with the psychiatrist turns out to be a *big* deal when it is the only deal in town. The relationship with the psychiatrist can give the lonely patient a taste of a committed companionship, albeit within professional boundaries.

The "space" between commitment and boundary, offering the patient a creative tension to work with, is called *the frame*.

＊

Not far from the frame between the psychiatrist and the patient is the frame between the parent and the child: the inability of the parent to fulfill every single wish of the child is a boundary, pushing the child to grow psychologically within the commitment of the parent. Growth happens through mastering the tension created by the external parent-child frame, as the child gets to experience both the limits of the relationship with the parent, and the commitment of the parent within the limits. Later, an internal frame, developed through learning, mirrors the external frame, allowing the child to efficiently master emotions not only related to the parent, but related to the surrounding environment in general. Thus, the frame does not only protect; it moves the child forward. It becomes the lens through which the child relates to the world.

Likewise, the frame between the psychiatrist and the patient does not only protect; it moves the patient forward. The frame is a catalyst for progress, giving direction to organize the patient-psychiatrist mind, which would otherwise fall in disarray. Later, an internal frame, developed through learning, mirrors the external frame, organizing the patient's brain not only in regards to the psychiatrist, but to the surrounding environment in general. Organizing the brain influences the predictability of emotions, and thus the degree of emotional self-control.

✳

Let's now look further into the internal frame within the brain. As a reminder, a brain has a top brain (the abstract brain, where cognitive rules are represented, including the conscious values), and an iguana brain—made of the middle brain (the emotional brain) and the bottom brain (the instinctual brain). The cognitive rules of the top brain represent both a commitment and the boundaries of the commitment. Commitment and boundaries form the internal frame of the brain, which cuts through the raw emotions of the iguana brain, pushing emotions to align, by adaptation, to the cognitive rules of the top brain.

The alignment of emotions to the cognitive rules shifts emotions from being raw in the iguana brain, to being adapted to the top brain. As the shift happens, the iguana brain splits functionally into the middle brain (the emotional brain) and the bottom brain (the instinctual brain). Since the middle brain (the emotional brain) now belongs to the top brain—by alignment of the emotions to cognitive rules— the iguana brain is not a functional entity anymore, leaving only the top brain—with emotions now—and the bottom brain—instinctual, and not accompanied by emotions anymore. The cognitive rules gain emotions, while instincts lose emotions.

Functionally speaking, the iguana brain has died, the *emotional* top brain was born, and the instinctual brain, while left alive, is not accompanied by emotions anymore. Structurally speaking, the iguana brain has split, the top brain is now connected to emotions, and the instinctual brain is now disconnected from emotions.

The newly expanded top brain, now including emotions, is an opportunity for self-control over emotions from the top, as opposed to a loss of emotional control to instincts. Suffering, which is an emotion, can come to an end with self-control over emotions. Note the difference between suffering and physical pain, an instinctual sensation connected to the body which drives the fight-or-flight survival response.

✳

We said earlier that the turning the other cheek faces resistance from antagonistic emotions, like anger. With the functional death of the iguana brain and the birth of the emotional top brain, the behavior of turning the other cheek can happen without the cost of negotiating against emotions, as now emotions are not lost to instincts anymore, and therefore not antagonistic anymore. The emotions are now aligned to cognitive rules, which include the conscious values—in this case, the value of turning the other cheek. This alignment makes negotiating against the emotions unnecessary. Instead, emotions, cognitive rules and behavior are all aligned now.

✻

There is a frame we haven't talked about yet: a frame originating a long time ago, playing a role ever since, universal in nature, and hidden from sight. It is the frame between God and the human being. The boundary of God's commitment was broken when the human being fell outside the frame, by eating the forbidden fruit.

One may wonder whether life on Earth is a big therapy session in which God, like a psychiatrist, listens to what goes on, does not talk back much, and allows growth to unfold within the mental space between the healer and the subject of healing. Within the boundary, a deep commitment is in place. Boundary and commitment form together the frame.

By breaking the boundary, the human being faces the limits of the commitment of God. Not everything is possible with God. The frame of the relationship with God is predictable, and therefore can be relied upon. On the contrary, evil does not play by rules. Instead, evil is unpredictable, and therefore unreliable. For the brain to be reliable, predictability is required. Thus, for the brain to become reliable, finding back the frame of the relationship with God is necessary.

On the way to finding back the frame, getting to know God is important, to delineate the commitment and the boundary. And thus, for the brain to become reliable, the Bible is necessary. Later, an internal frame, developed through learning, mirrors the external frame of the relationship with God, becoming the lens through which the human being relates not only to God, but the world in general. It shapes not only the behavior toward God, but toward the surrounding environment too. Keeping on this lens leads to predictability of emotions, which makes emotions reliable, thus paving the way to self-control over the behavior.

*

Adam and Eve did not know of their own nakedness at first. Nakedness stands here as a metaphor for inner emptiness, manifested as lack of emotional self-control—leaving Adam and Eve filled with the trust in God instead. Adam and Eve did not see their lack of emotional self-control: the trust in God was "blind".

Without emotional self-control, Adam and Eve felt an emotional change at the time when the serpent asked them to replace the trust in God with mistrust in God. The emotional tie to God was pushed away, making room for an emotional disconnect from God. In the absence of the emotional tie to God, the behavior followed the emotional disconnect from God: Adam and Eve acted for the serpent, not for God, when eating the forbidden fruit of knowledge of good *and* evil.

Knowledge is information, which in turn has an impact on the body in the form of an emotion. The newly gained information, of the difference between good *and* evil, has an impact on the body in the form of the emotion of pain.

"Your eyes shall be opened", said the serpent. Now their eyes were open. First and foremost, the eyes were emotional eyes, "seeing" through feeling emotional pain the difference between good *and* evil. The Bible in the Book of Genesis mentions the "painful toil" of eating food from the ground and the "pains in childbearing". And thus the suffering of the two labors of life arrived: the labor of maintaining life, and the labor of bringing a new life into the world.

After Adam and Eve became painfully aware of their lack of emotional self-control, Adam and Eve hid from God. Then, God made a cover of garments of skin for them. The cover of garments of skin symbolizes the hiding under the unconscious of the awareness of lack of emotional self-control. By pushing the lack of emotional self-control outside awareness, the unconscious stops the emotional pain (of the knowledge of the difference between good *and* evil) from being felt, allowing Adam and Eve to relate to God once again. Their "eyes" which saw the emotional pain are closed now.

But great tension is still present nowadays inside the human being: psychoanalysis describes the unbearable tension between the unconscious and conscious, making the unconscious necessary to hide the emotional pain within the dark side of the human. Well-described mechanisms of defense provide the garment between the unconscious and the conscious. Through the tension in the garment, the conscious creation of the human by way of choice waits to happen—via emotional self-control, against the depth of the unconscious. The choice is of the conscious values—about which psychoanalysis stays mostly silent. As for the human will, the lack of emotional self-control dilutes the human will, making it less than free due to out-of-control emotions.

Note that the lack of emotional self-control is not a result of the unconscious, but of the impossibility, even for God, to create a human being that is both free and *simultaneously* imposed upon what to feel emotions for. Gaining emotional self-control requires *learning in time* to overcome through trust emotions led by mistrust, which ultimately leads to the development of trust-based emotions instead of mistrust-based emotions. And so, time arrived in the world. As for the will, emotional self-control frees up the will from out-of-control emotions.

After God saw in time that the knowledge of good *and* evil prevented the human beings from overcoming through trust the emotions led by mistrust, and thus prevented the human beings from gaining emotional self-control, God allowed Christ to become the stepping-stone of emotional self-control. The trust in God against the knowledge of good *and* evil allowed Christ to have emotional self-control. This way, the human beings are offered the opportunity of trust in something less abstract than a God behind the scenes: trust in a God that took the form of a human being, to which other human beings can relate without the effort to figure out what God is means. Christ showed what God means.

When following in Christ's footsteps, a human being aims for emotional self-control. This puts the human being closer to God—against the still present force of the unconscious. Full awareness, once achieved, becomes not of the mistrust due to lack of emotional self-control anymore, but of the image of Christ within, leading to self-control over emotions.

When emotional self-control is gained through the image of Christ within, the garment hiding the awareness of lack of emotional self-control under the unconscious becomes useless. With nothing left to hide, the unconscious becomes unnecessary. Consequently, the tension with God disappears, allowing the human being to relate to God without the need for an unconscious anymore. Subsequently, the Garden of Eden becomes unnecessary, as the beauty of human life comes now from within, sustaining foreverness in relation with God without needing an outside tree of life for it. What was a previous inner emptiness that allowed the knowledge of good *and* evil is now filled with the knowledge of good *against* evil.

This new knowledge is information, which impacts the body in the form of an emotion that is pain-free, since the discrepancy between good *and* evil was replaced by the consistency of the good *against* evil. The human being is not ambivalent anymore. Instead, in full awareness of good *and* evil, the human being makes a choice, in full emotional control, for good *against* evil.

DNA

Genetics are a driver of the unconscious. Example: the genetics of reproduction drive the unconscious thoughts of intimacy.

Imagine the genes like a piece of Swiss cheese, with the usual holes in it. The holes of the Swiss cheese give directions to the unconscious thoughts: when electrical current is discharged through the holes, thoughts of unconscious nature take place. Note that the Swiss cheese, the genes, exist even in the absence of the discharge of electrical current, in the absence of the unconscious thoughts.

A battle takes place between two memories: the genetic DNA and the memory of values. Proteins make up the genetic DNA, and proteins store values in the brain. The same way DNA exists independent of the unconscious thoughts it drives, the memory of values does not depend on the thoughts about values.

The hardware made up of proteins operates by software: the information. Thus, at the table of the ego function, the genetic DNA software and the value-based software negotiate. As noted before, values come to the brain by way of the mind.

Let's use the Christian model, to illustrate: software number 1, the genetic DNA software; software number 2, the values driven by the mind formed between the human and God. The ego function is called upon to negotiate, within each person, between the two softwares: one is of a mindless structure—the brain proteins that make up the genetic DNA—and the other is of a mindful structure—the brain proteins that memorize the values driven by the mind between the human and God.

Note that structure and function are not mutually exclusive, despite the genetic DNA being a broken structure by virtue of a broken function limited to structure replication. The goal of the mind formed between God and the human is not so much to infinitely replicate broken structures, but to unleash the functional potential within each structure—by infusing the broken genetic DNA with values driven by the mind between God and the human. This creates a mindful structure—an advanced form of structure when compared to the mindless structure of a genetic-driven body.

As noted before, after the forbidden fruit was eaten, the human beings lost access to the tree of life. This brings up the issue of the necessity of death. The knowledge of the difference between good *and* evil damaged the genetic DNA, which now uses the human ego to press against the mind formed between God and the human, for the mind to snap. Thus, the broken genetic DNA becomes a form of ballast, a drag, hosting destruction through unpredictability, and thus unreliability. Consequently, the broken genetic DNA lacks sustainability over time. In the broken genetic DNA, death is imprinted.

For death to go away, the broken genetic DNA material has to go away. As such, the Bible says that the way to the tree of life has been blocked by a cherubim—in order for the broken genetic DNA to not access forever life. Note the logical impediment: for death to go away, the access to forever life was blocked. To harmonize this apparent logical hiccup, let's note that death here stands for functional death. For a malfunction to not be everlasting, the structure that sustains the malfunction has to go away, freeing the malfunction to return to good function.

God has foreworn the human being before the fall (Genesis 2:14 "when you eat from it you will certainly die"). The serpent, on the other side, disputed it (Genesis 3:4 "you will not certainly die"). The dispute by the serpent tempted the human being to break the boundary of God's commitment, which ultimately led to death: first of the function in relation with God (the mind human-God), then of the human structure (the human body).

The necessity of death sustains creation, in the hope that the mind formed with God by the human on Earth becomes a blueprint of a reformed genetic DNA material for a new life. Then, the mind formed with God drives the reformed genetic DNA material, as opposed to the previous genetic DNA material being broken by the fall away from God. To become a blueprint of a reformed genetic DNA material for a new life, the mind formed with God by the human on Earth targets self-control over emotions, to replace the inner emptiness of lack of emotional self-control.

Note that, as it stands now, the unconscious is in the brain—but is not detectable easily by the conscious part of the brain. Instead, the unconscious tends to first manifest itself in an organ different than the brain, despite the location of the unconscious being in the brain. The unconscious tends to first manifest itself in the heart. The unconscious emotion (the impact of the unconscious information on the body) gives the first signal of its existence from the unconscious brain to the conscious brain by increasing the heart rate, an expression of the unconscious emotion noticeable by the conscious part of the brain.

Case in point: two people not consciously attracted to each other find their hearts beating faster as the conversation goes on, to their surprise. Consciously, they had no idea. Unconsciously, they did have an idea. But because the idea was unconscious, they did not have an idea that they had an idea. It was only their hearts that gave up the unconscious idea, through increased heart rates.

Note how the unconscious is not present in the abstract: the unconscious is present in a network of neurons in the brain, outside of brain awareness. But the network of neurons is real, despite being unconscious. Consequently, the unconscious is not only a *function* parallel in the brain to the conscious function. The unconscious is also a *structure*, in the form of a network of neurons, parallel to the conscious structure of the brain.

With two parallel structures in the brain (the unconscious and the conscious) having two parallel functions (the unconscious and, respectively, the conscious), self-control by the human over the entire brain is missing. With the unconscious emotions deterring self-control over the brain, the mind formed with God is at serious peril if depending on the human lone. Instead, an intervention from God is necessary.

And so, where the human fails, God raises to the occasion. Or, as noted by Saint Paul the Apostle in Romans 5:20, where sin abounds, grace abounds more.

The mind formed with God is protected from unconscious emotions by the grace of God through the separation between the conscious and the unconscious (the "garment"), putting the mind formed with God in the position to target emotional self-control, in order to replace the inner emptiness of lack of emotional self-control (the "nakedness"). The grace of God is a necessary ingredient to achieve emotional self-control against the depth of the unconscious.

DIAGNOSIS

In an attempt to functionally deal with the unconscious structure of the brain, an apparent roadblock for the conscious structure of the brain is brain illness. This raises the issue of how to get through the roadblock of brain illness. The fight against brain illness involves putting the correct diagnosis in order to guide the treatment selection effectively, since not every treatment works for every brain illness.

Unfortunately, on the way to putting the correct diagnosis, perils pop up from time to time.

In the field of psychiatry, usually there is more to the story than what the patient says. When compared to a checklist of symptoms, listening to the story is important, as inferring from the story may be necessary, to sort out the line between what is going on and what the patient believes is going on. Beyond the richness of every story told by the patient may hide a fracture in the narrative, which brain illness brings about.

Due to the subjective nature of psychiatric complaints, and the paucity of objective testing, critical thinking is important in psychiatry relative to other specialties. Scrutinizing the patient with the use of critical thinking stands the chance to increase the accuracy of the diagnosis. On the other hand, bypassing critical thinking on behalf of being led unconditionally by a patient gets the psychiatrist to where the patient wants, but not necessarily to an accurate diagnosis.

＊

Behind every narrow question looms a large answer that goes beyond what the question asks for. For instance, when the psychiatrist asks for the personal educational history, the answer tells more about the patient than the question asks for.

With the experience of seeing patients over time, the psychiatrist can recognize sharp turns in the story, when listening to the answer to the education question; for example, a pause followed by a sigh in the patient's response is a good clue of a hidden emotion underneath the pause, pointing to a loaded experience as opposed to a routine, forgettable event. Chances are that something went wrong at the recollected moment in time—maybe with important relationships in childhood (parents, caregivers, teachers, or the like), or possibly even an abuse took place. The clue of the pause followed by a sigh is not the answer. It is an invitation to explore in-depth what happened that build up to the sigh after the pause.

Asking about the patient's education is an opportunity for the psychiatrist to infer information by the way the patient breathes, the choice of words, the level of eye contact, the change in affect, the general openness of the conversation. The psychiatrist can sometimes pick up leads based on the tension exuded, the sense of fear, the speed of dialogue.

In general, when the psychiatrist asks the patient about a narrow topic, the psychiatrist opens a door to the history of the patient beyond the narrow topic. The "music" played by the patient in answering points to the twists and turns in the story, past what the question asks for.

※

Brain illness could cause the patient to not tell the truth. This may not be a lie, but a symptom that can mask the true diagnosis. For instance:

1. not telling the truth can be a psychological denial mechanism of non-autonomous nature, automatic, not within self-control, with the role of masking the otherwise unbearable emotional pain of facing the truth. An example is of a patient with multiple personality disorder;
2. not telling the truth can be a delusional thinking about what the patient believes the truth is. An example is of a patient with schizophrenia;
3. not telling the truth can be a need to be taken care of that is non-autonomous, automatic, not within self-control, illness-driven. An example is of a patient with factitious disorder.

Some patients do not tell the truth for a well-defined reason, autonomous, within self-control, choice-driven, not illness-driven. For instance:

1. A patient with a mild depression that lies about hearing voices to elevate the diagnosis from minor depression to major depression with psychotic features, in order to increase the chance of getting disability benefits;
2. A patient with generalized anxiety disorder that lies about being in pain, in order to obtain a narcotics prescription;
3. A patient with antisocial personality disorder that lies about being homeless, in order to qualify for a housing program subsidized for the homeless.

When choosing to lie, a patient likely starts by anxiously scrutinizing how the psychiatrist reacts, with an apparent expectation of validation. Once convinced that the psychiatrist bought the lie, the patient likely relaxes, exhibiting a release of tension. This is in contrast with telling a true story from the patient's point of view, where the patient likely becomes progressively tenser as the emotions build up during the retelling. Here the variation in tension likely follows the story turns consistently, while for a patient who chooses to lie, the tension likely does not synchronize with the story turns.

A patient may choose to lie more sophisticatedly than by simply making stuff up from the scratch: a patient may choose to lie by playing up symptoms through exaggeration. Carefulness is necessary, because underneath the exaggeration may hide a true diagnosis, different than the diagnosis that the patient is trying to portray. An appearance of lying does not automatically eliminate the possibility of a true diagnosis.

✳

When the psychiatrist faces a lie from the patient, piercing the lie cannot always be accomplished with honey. The legal establishment knows this well. That's why the legal establishment gives to the police inherently coercive techniques of interrogation. Without the inherently coercive techniques, many criminals would never be successfully prosecuted. Due to ethical concerns, the psychiatrist is not given the inherently coercive techniques of interrogation like a cop, despite the fact that a patient does not always tell the truth.

Left only with confronting the patient about lying, what happens when the psychiatrist does so? To start with, the patient's satisfaction can go south, right away. The patient might quit treatment. Or, the patient might make noise, by bringing complaints, asking for investigations, or even threatening with lawyers. A lawsuit, even frivolous, can damage the reputation of the psychiatrist, by forcing a "yes" answer to a credentialing question from healthcare organizations "Have you ever been sued by a patient?"

Because confrontation can strain the relationship with the patient, and honey does not always pierce a lie, and inherently coercive techniques of interrogation are not on the table, the psychiatrist is tempted to simply go along with what the patient says. This becomes particularly concerning when assessing risk.

✳

Assessing risk is integral to the psychiatric evaluation. An example is evaluating the risk of harm.

Just because a patient says "Doc, I am harmful" does not always mean the patient is harmful. When not looking beyond words, the psychiatrist has a quick fix: admit to the hospital every patient that says the password "Doc, I am harmful", without performing an additional risk assessment. In doing so, the psychiatrist replaces the accuracy of the risk assessment with the cost of the hospitalization.

An opposite example is of a patient who says "Doc, there is nothing wrong with me", while the family, neighbors and the outpatient social worker report the patient is at serious risk of harm. When disregarding the collateral information and discharging the patient from the ER on the words of the patient only, the psychiatrist replaces the accuracy of the risk assessment with the cost ignoring the risk of harm.

As seen in these examples, the words of the patient, like a party statements in a legal proceeding, amount to less than the full weight of the evidence. Beyond the words of the patient is the formulation of the clinical theory of the case, also known as the clinical formulation, which looks at where the story goes *despite* what the patient says. The clinical formulation may point to a different direction than what the patient says. Thus, to mitigate the risk of being inaccurate, an assessment in psychiatry needs to go beyond the patient's words.

Formulating the clinical theory of the case means inferring a professional impression based on clinical reasoning beyond factual information. Said in simpler terms, by merely listening to the facts as stated by the patient, the psychiatrist has not yet discovered all the pieces of the puzzle.

Nonetheless, while listening to the facts stated by the patient, the psychiatrist can still form a theory of where the puzzle is going. Then, the psychiatrist can compare the subsequently found pieces of the puzzle with the anticipated puzzle, to determine if the theory holds. As the psychiatrist finds new pieces of the puzzle, the number of possibilities of where the puzzle is going decreases, and the puzzle becomes more predictable. As even more pieces of the puzzle are discovered, the puzzle—the narrative diagnosis—becomes clear.

✳

Unfortunately, the system does not incentivize a psychiatrist to formulate a clinical theory of the case in order to clarify the narrative diagnosis. Instead, the system allows for a dry checklist of symptoms to drive a cookie-cutter diagnosis, which can be far from the reality of a narrative diagnosis discovered by employing a clinical formulation.

Consider the example of a patient sent in for a psychiatric evaluation by a third party and too shy to report symptoms. Afraid to open up, the patient often denies a problem exists. Without a clinical theory of the case, the "diagnosis" based only on the words of the patient is nothing: "Doc, I have no complaints". But in the differential diagnosis can hide, for instance, PTSD. The too shy patient may actually have a psychologically defensive posture, being on the lookout for the next trauma, which interferes with functioning in an undisclosed manner—"Doc, I have no complaints".

The way to correctly diagnose a patient like this is by detecting a pattern from milestones—which can include hospitalizations, jail time, suicidal behavior, or incidents of aggression—as well as from third party information, documents, and even non-verbal cues, such as the inflections in the patient's own voice, which can each point to the difference between what the patient says and what goes on in reality. Subsequently, the pattern analysis may validate the clinical theory of the case, that the patient has PTSD despite saying "Doc, I have no complaints". Or, the pattern analysis may invalidate the clinical theory of the case, validating instead what the patient says, that nothing is wrong beyond the shyness itself.

Let's now consider the example of the patient who wants disability benefits no matter what. The patient might recite to the psychiatrist the diagnostic criteria of an illness. Legend has it that some attorneys give disability applicants "cheat sheets", to memorize the elements of the definition of brain illness from the psychiatric manual. The patient is caught between two opposing forces: the doctor who wants the patient better, and the disability system that sets the minimum threshold for how ill the patient must be in order to collect money.

Practically speaking, the patient is left between getting healthy for no pay or pursuing sickness for pay.

When wanting disability benefits no matter what, the patient may report a dry checklist of symptoms, hoping that the psychiatrist will arrive at a convenient diagnosis. But the psychiatrist could go to the length of formulating a clinical theory of the case, instead of merely going by a dry checklist of symptoms. Faced with an emotionless verbal report from the patient, word-by-word from the book, with no nuances and no variation from the norm, the psychiatrist is looking at the hypothesis that "the patient is lying".

Next, the psychiatrist gathers more pieces of the puzzle, while observing whether the clinical theory of the case holds (that "the patient is lying") or is contradicted by the newly found evidence. Once the psychiatrist has enough pieces of the puzzle, an ongoing adjustment of the clinical theory of the case paves the way to a more certain picture—for instance, a diagnosis of malingering, instead of a mere clinical theory that "the patient is lying".

Let's now consider some examples of the lifesaving difference between a clinical theory of the case and a dry checklist. The patient complains of a panic attack and it turns out to be a walking heart attack. The patient acts drunk and it turns out to be a walking stroke. The patient acts psychotic and it turns out to be a ruptured brain aneurysm. The patient denies suicidal ideations and it turns out an overdose just happened, and a goodbye letter. In each of these examples, the life of the patient depends on the quick critical thinking of the psychiatrist to formulate a clinical theory of the case, beyond a misleading dry checklist.

A good way to diagnose is to consider more than what the patient says: to consider telltale signs beyond what the patient says, and the mechanisms involved in the difference between the telltale signs and what the patient says. The mechanisms of the difference lead to the causes of the difference, improving the accuracy of the diagnosis.

A mere checklist of what the patient says can result in an inaccurate diagnosis. Moreover, a checklist alone can kill, even if only because to capture on it everything that can possibly go wrong is impractical. The key to an accurate diagnosis lies many times in the subtle difference in patterns that checklists simply don't have enough depth to capture.

﹡

The formulation of the clinical theory of the case in psychiatry beyond checklists is the equivalent of the theory of the case in law. How simple is to formulate a clinical theory of a case?

1. First, the psychiatrist needs to help the patient verbalize what's going on, a task in itself not to be taken for granted; then, the psychiatrist needs to dig out clinical bits of information hidden beyond the surface of the patient's words—found concealed in patient's actions beyond words, and found collaterally too, like in medical records and third-party statements;

2. Subsequently, there is a need to articulate a coherence between the seemingly disparate bits of information gathered;

3. Next, there is a need to interpret clinical bits of information through diagnostic criteria and through clinical narrative patterns learned in psychiatric training—this is not a layperson's giving of a street opinion on what's happening;

4. In addition, the psychiatrist has to withstand the temptation to become defensive against non-clinical forces pushing against the autonomy of the psychiatrist (e.g. insurance companies, employers, governmental agencies, lawyers);

5. Plus, the psychiatrist has to balance the values of the psychiatrist (e.g. that suffering can be an integral part of normal

growth) versus the values of the patient (e.g. that suffering is always abnormal) within a value-driven parallel psychiatry.

How to balance within a value-driven parallel psychiatry the values of the psychiatrist with the values of the patient, for the purpose of formulating the clinical theory of the case, is a function of what parallel psychiatry we are in, which can evolve dynamically based on how patient is doing. For instance, a very ill patient requires, for treatment adherence purpose, a value-driven parallel psychiatry consistent with the patient's values. As the patient gets better, to update the clinical formulation, the psychiatrist can softly slide to a value-driven psychiatry closer to the psychiatrist's values, of which the patient can opt out. In light of the patient's needs as seen by the psychiatrist, a frank discussion can take place, when the patient is ready, about the differences in values between the patient and the psychiatrist.

Let's expand a little bit on the non-clinical forces pushing against the autonomy of the psychiatrist. Insurance companies may deny payment for medically necessary treatment. Employers may press for customer satisfaction. Governmental agencies are ready to investigate complaints from unhappy patients. Lawyers are ready to sue on use of private information to gather collateral information. The psychiatrist prefers insurance companies to pay, employers to be happy, governmental agencies to be silent, and lawyers to be kept at bay. All these preferences can conflict with the accuracy of the clinical theory of the case, which can inadvertently turn into an insurance theory of the case, an employer theory of the case, a governmental theory of the case, or a legal theory of the case.

Therefore, formulating a clinical theory of the case is not a simple matter, but a high-end executive function of the psychiatrist's brain. It requires a solid wealth of professional knowledge of diagnostic criteria and clinical narrative patterns, to reflect on seemingly disparate clinical bits of information, some pulled from beyond the patient's words, and it requires to maintain the autonomy of the psychiatrist within a value-driven parallel psychiatry. Formulating the clinical theory of the case requires the psychiatrist to battle on two fronts: one within psychiatry (to determine what's going on with the patient) and one outside psychiatry (to fight off non-clinical forces).

✳

During the formulation of the clinical theory of the case, the evidence is not always crystal clear (direct evidence). Sometimes, the evidence is less than crystal clear (circumstantial evidence), requiring inference, or corroboration with more evidence. In psychiatric training, little emphasis is placed on overcoming the lack of direct evidence with circumstantial evidence.

Here is an example of direct evidence: "Doc, when the cops showed up I was breaking into my neighbor's house with an ax. I wanted to kill my neighbor". This statement has on its face the intent of the patient.

Here is an example of circumstantial evidence: the patient was picked up by the police while trying to break into the neighbor's house with an ax; the patient's spouse said that the patient had an argument with the neighbor earlier in the day, and that the patient stopped his psychiatric medications three days ago on his own. It is not clear from the circumstantial evidence that the patient was about to kill the neighbor, but one might infer it. Let's suppose that the patient follows up by saying, "There is nothing wrong with me, Doc" and offers no *direct* evidence of anything being wrong. Still, the *circumstantial* evidence can be strong enough to *infer* from it the seriousness of the matter.

The accuracy of the clinical theory of the case can increase when the circumstantial evidence is taken in account. In some cases, the circumstantial evidence opposes the direct evidence, and may even overturn it.

✳

Psychiatrists don't always give the circumstantial evidence the required weight. It may have to do with the consumerist approach: "Patient, what's bothering you? That? Okay, here's a pill for it".

Training in critical thinking, like a detective, can be useful to a psychiatrist, as a way to look beyond the surface of the patient's words, to find additional evidence, whether direct or circumstantial in nature. Then, the newly found evidence needs to be articulated, along with what the patient says, into the clinical theory of the case, which spells out the opinion of the psychiatrist. This opinion has the weight of an actual fact on which the law can be applied.

To put it more clearly: the opinion of the psychiatrist, spelled out in the clinical theory of the case, *is a fact* as far as the law is concerned. Without the clinical theory of the case, an important fact is missing: the opinion of the psychiatrist, on which the law applies. By wrongly applying the law to facts inaccurate by omission, the patient's rights can be inadvertently violated. An example comes from a jurisdiction where the law permits the involuntary hospital admission of a patient only if three facts are established: 1. The patient acts dangerously; 2. The patient has a brain illness; and 3. The patient acts dangerously *due to* the brain illness.

Let's assume that facts number 1 and 2 have already been established: that the patient acts dangerously, and that the patient has a brain illness. However, without the opinion of the psychiatrist, the fact number 3 cannot be established by direct or circumstantial evidence, because a layperson cannot establish the causal link between the dangerous behavior and the brain illness, since a layperson doesn't have the professional knowledge of when to deem the behavior illness-driven, non-autonomous, automatic, outside self-control—as opposed to choice-driven, autonomous, not automatic, within self-control.

To establish the causal connection (fact number 3), the clinical theory of the case spelling out the professional opinion of the psychiatrist is required. Only by establishing fact 3, in addition to the already established facts 1 and 2, is the legal threshold met for the involuntary hospital admission in that jurisdiction.

Without establishing fact 3 (the causal connection), the mere establishing of fact 1 (the dangerous behavior) and fact 2 (the brain illness) falls short of the legal threshold in the jurisdiction that requires facts 1, 2 *and* 3 for involuntary hospital admission. Admitting the patient to the hospital involuntarily merely on facts 1 and 2 violates the patient's right of freedom from hospitalization in that jurisdiction. An involuntary hospital admission in the absence of establishing a causal connection between the dangerous behavior and brain illness is then unlawful. Because the causal connection stems from the professional opinion of the psychiatrist articulated in the clinical theory of the case, skipping the clinical theory of the case becomes unlawful in that jurisdiction.

Interestingly, the psychiatrist is called here to make a legal judgment dressed as a medical decision: whether a legal standard is met, spelled out in the letter of the law for involuntary hospital admission.

✴

To find justice, the legal process has three different people filling three roles of, respectively, defense attorney, prosecutor, and judge. In comparison, to diagnose accurately, the psychiatric process has one person filling three roles simultaneously: the psychiatrist is the caregiver (the equivalent of the defense attorney), the lie detector (the equivalent of the prosecutor), and the decision-maker (the equivalent of the judge).

A judge that takes the job of defense attorney while continuing to be the judge on the same case is tempted to judge for the defense. Or, a judge that takes the job of prosecutor while continuing to be the judge on the same case is tempted to judge for the prosecution. Likewise, a psychiatrist simultaneously filling the roles of caregiver, lie detector and decision-maker is tempted to diagnose according to what matters most to the psychiatrist—like the predominant value of the psychiatrist; this can be, for instance, making money (e.g. when insurance companies pay only for certain diagnoses), or having free time (e.g. when arriving to one diagnosis is faster than to another), or defensiveness against lawsuits (e.g. when a diagnosis covers a wider range of differential diagnoses than another).

Having three different people fill three roles in the legal process searching for justice is a checks and balances mechanism. The same checks and balances mechanism is absent in the psychiatric process, which searches for the diagnosis with the same psychiatrist in three roles at once. The psychiatric process relies on the moral strength of the psychiatrist to come up with a diagnosis that is supposed to be accurate.

To resolve the tension between what matters most to the psychiatrist and the need for accuracy in diagnosis, the psychiatric process might consider an alternative to relying on the moral strength of the psychiatrist: to spread the three simultaneous roles of the psychiatrist (caregiver, lie detector, decision-maker) to more than one person.

That's a nice dream for now.

DREAMS

In general, people are encouraged to follow dreams. That being said, getting stuck in an impossible dream can lead to brain illness. Adjustment to reality is better than chasing an impossible dream.

Let's take the example of a person who dreams of becoming a successful professional dancer. "You can do it". "You've got the power inside you". "Just try harder". All of these may be good for raising morale, but the majority of people cannot come close to being successful professional dancers. So, the person either adjusts the dream to reality, or suffers from not being able to fulfill a too large dream.

When the suffering of not adjusting the dream to reality lasts, brain illness is invited into the picture. When the dreamer, however, is able to adjust the dream from being a successful professional dancer to, let's say, being part of a nonprofessional, but easy to enjoy dance club, then the individual stands a much better chance of fulfilling the dream—adjusted to reality this time. Instead of the dream controlling the dreamer, the dreamer now controls the dream.

Adjusting the dream to reality can be protective of the brain health of the dreamer. For people who cannot get out of an impossible dream on their own, therapy can provide a way out. Through therapy, the unadjustable impossible dream is brought out into the open, under the light of reflection. Gaining therapy-driven insight into the underpinnings of the unadjustable impossible dream allows for tackling the underpinnings—offering a reality check that challenges the underpinnings, with the purpose of ultimately adjusting the dream to what is possible.

✳

At times, not even therapy can adjust the seemingly impossible dream. Left without options, the patient wonders what to do, alone. When all chances to get out of the dream seem lost, the right partner may come along. The right partner perceives the patient's dancing to be beautiful, which in turn brings an ease to how the patient dances.

When not possible alone, beauty can be attained in pairs. The reference system is the key: while the mind between the audience and the dancer would be harsh on the dancer, crashing the dancer into the floor, the mind between the right partner and the dancer is easy on the dancer, lifting the dancer to a smile again.

A good relationship changes reality. What seemed impossible at first turns real. With the right partner, a seeming misfit can change dramatically. A dream once impossible to fulfill stands the chance to be fulfilled now: an otherwise bad dancer can truly become a star in a reality made of two people, where no other reference point is present but the right partner in dancing. Both may be bad dancers, but nonetheless, each of them perceives the other as dancing wonderfully together.

*

To retreat in dreams is not always a conscious choice. Instead, retreating in dreams can happen automatically. An example is dissociative fugue, where the patient is half asleep, half awake. In dissociative fugue, the patient literally takes off to a different geographical location, without being aware of the travel until arriving to the destination.

Does the patient mean to go to the destination, or is the illness controlling the travel? Unfortunately, in dissociative fugue, the illness controls the travel. The patient has no intent of going there. Because the patient is unable to consciously tolerate the emotional pain, the unconscious takes over and does the travel for the patient, moving the patient away from the geographical location associated with emotional pain.

While in dissociative fugue the patient is half asleep, half awake, other patients dream fully asleep—like a victim of abuse retreating into a dream about the time before the abuse, when things were well. And yet, other patients dream fully awake—like a person with borderline personality disorder that dreams about one day being a famous artist.

A seemingly good but unattainable dream raises a tension within the patient: on one side the patient likes the dream, since the dream appears to be good, on the other side the patient hates the dream, since the dream is unattainable. When unable to resolve the tension, the patient gets stuck in the ambivalence of simultaneously loving and hating the dream. Like a moving pendulum, the patient stuck in the ambivalence embraces the seemingly good dream, only to find the dream unattainable, followed by the patient retreating to build momentum for chasing the dream again, unattainable once more.

The pendulum moving back and forth is a failure to fulfill the dream and at the same time a failure to adjust the dream to reality. This opens the door to brain illness, by neurotically chasing the dream—back and forth, repeatedly, to no avail. A relentless neurotic chase of the dream has a traumatic impact on the dreamer.

TRAUMA

A traumatized patient once said, "I'm angry with God. He doesn't care about me anymore". What can be done about this? How to deal with the loss of faith secondary to trauma?

When trauma occurs, new information comes to the body, met by the two softwares discussed earlier: the software of the conscious values and the genetic DNA software. The ego function negotiates between the two softwares on how to integrate the new information, about trauma. When either the conscious values software is strong (dissipating the trauma) or the genetic DNA software is strong (dissipating the trauma), then the new information is kept at bay and does not modify the unconscious of the victim.

The conscious values software can dissipate trauma the same way a muscle dissipates a physical impact: by strength. Training is required—for the muscle, in the physical environment; for the conscious values software, in the mental environment. For instance, a long time ago, people in villages said after an adversity: "It was God's wish". Whether it was in fact God's wish was not known for sure. But people assumed so, by training in the mental environment of the close-knit village: their conscious values software was trained to pivot on the value "It was God's wish", thus maintaining a positive outlook despite adversity, and absorbing the trauma impact like a shield.

Without training to dissipate the trauma, the conscious values software loses the strength of a line of defense. Subsequently, complications from trauma are more likely to develop, such as a victim mentality—"I am a victim of what didn't happen yet."

As trauma pierces through the conscious values software, trauma gets in the position to attack the next line of defense, the genetic DNA software. There, a strong genetic line of defense resists against trauma, but a vulnerable genetic line of defense allows trauma to penetrate deep, in the unconscious of the victim, where trauma can modify the unconscious.

The modification of the unconscious by the trauma can happen through an ego function fracture that disconnects the conscious check from the unconscious. Consequently, conscious values are severed from the unconscious thoughts, which in turn has a physical impact on the body in the form of the unhinged emotion of fear (in general, emotions do not happen out of the blue, being the physical impact of information on the body; they can be noticed, for instance, as an increase in the heart beat).

While the physical impact of the traumatic information on the body is the emotion of fear, note the difference between fear and unhinged fear. Preceding the reaction of fight or flight, fear is a survival mechanism after trauma, that goes away after the fight or flight. Unhinged fear does not go away. Consequently, unhinged fear can result in manifestations with recurrent character, such as: avoidant behavior, nightmares, flashbacks, intrusive thoughts, hypervigilance. When these recurrent manifestations interfere with functioning, the illness of PTSD arrives.

In PTSD, an imbalance occurs between the conscious values and the unconscious thoughts, with the unconscious thoughts gaining momentum against the conscious values. Moreover, the traumatic event is relived through the symptoms of PTSD, outside the choice of the victim, leading to re-traumatization by way of the symptoms. Then, the illness grasps with force the core of the victim, and tends to not go away easily. Consequently, PTSD becomes self-propelled through relieving the trauma by way of the symptoms. In addition, PTSD can be like a powder keg—ready to amplify the impact of a future trauma.

Previous to PTSD, the ego function trusted the conscious values against the unconscious thoughts. After the onset of PTSD, the fracture in ego function severs the trust in the conscious values against unconscious thoughts. Consequently, the unconscious thoughts are galvanized by the fracture in the ego function, to the point where "I do not trust" gets to reign in the unconscious. This is a critical unconscious change in PTSD—from "I trust" (under the influence of conscious values, through the ego function) to "I do not trust" (outside of the influence of conscious values, due to the fracture in ego function).

Nobody can control the unconscious directly. But people control it indirectly, through the ego function trusting conscious values. With the ego function fractured in PTSD, indirect control—the only control *over* the unconscious—is lost. Consequently, control shifts *to* the unconscious. In PTSD, the unconscious controls the victim, making

PTSD non-autonomous, not in self-control, automatic, and therefore an illness (as opposed to malingering—autonomous, in self-control, by choice, not an illness despite being a diagnostic category).

✳

Being aware of trauma is not required to be a victim of trauma. Even without awareness, trauma can damage the quality of the victim's life.

For instance, a child conditionally loved by the parent based on the school performance is traumatized by the absence of love when the school performance is less than expected. The traumatizing of the child happens even when the child is unaware of the trauma and believes instead the issue to be the school performance. To avoid traumatizing the child, the parent has no choice but to unconditionally preserve the full strength of the parent-child mind, which only love can give. This makes unconditional love necessary for trauma to be absent, and thus for the child to develop well. Conditional love is not enough, despite the child's unawareness of the trauma after a less than expected school performance.

In general, how aware a victim is of the trauma influences whether the victim facilitates avoidance of trauma, or repetition of trauma:

With awareness of trauma, fear of trauma is present, facilitating avoidance of trauma.

Without awareness of trauma, fear of trauma is not present, repetition of trauma is not guarded against, thus being facilitated, and can become habit forming (the patient doesn't know any better).

Who would have thought that conditional love based on school performance can result in PTSD?

✳

Trauma does not have to originate external to the victim. Sometimes trauma originates within the victim. For instance, a soldier hurting innocents in war can be traumatic to the soldier too, not only to the innocents hurt by the soldier. The conscious values software of the soldier enters in conflict with the incoming information about the soldier's behavior of hurting innocents, showing that a wrong has been committed: a "sin", or an error relative to the conscious values software of the soldier, which puts the soldier's trust in oneself in jeopardy.

The conflict of software nature between the conscious values and the incoming information about behavior *is the trauma* to the soldier. When the conflict is not resolved, PTSD knocks at the door. To resolve the conflict, either the conscious values of the soldier have to align with the behavior of hurting innocents (an undesirable outcome from the societal perspective), or the incoming information about behavior has to change (since information about old behavior cannot change retroactively, new behavior of compensatory nature is necessary—like doing acts of good deeds).

Resolving the conflict by overextending psychiatry from dealing with illnesses to therapizing away values can make for a dangerous society that gives up on values. Instead, resolving the conflict outside psychiatry by new behavior of compensatory nature, like doing acts of good deeds, presents itself as an opportunity. The incoming information about the new behavior of compensatory nature brings the overall incoming information about behavior closer to values (packaging the new behavior with the old behavior), thus decreasing the conflict between the overall behavior and values. Consequently, because the conflict is the trauma, a decrease in the conflict results in a reduction of trauma, and thus in lowering the risk of PTSD.

✳

Compassionate people surrounding the victim may not easily notice how the lack of trust due to trauma can take over the victim of trauma with PTSD. Then, the lack of trust due to trauma can generalize the behaviors of the victim with PTSD, colored by the automatisms in behavior from PTSD, toward people other than suspected aggressors in the future.

The automatisms in behavior from PTSD can be non-aggressive—like shyness, avoidance—or aggressive—like verbal outbursts, and even physical fighting through an increased fight-or-flight response.

When aggressive, the automatisms in behavior from PTSD can lead to a role reversal, turning the victim from being traumatized into traumatizing another person. Trauma then becomes contagious, contaminating others. Consequently, through trauma contagion, PTSD can be transmitted from one person to another.

The transmission of a brain illness is not the illness itself, just like the transmission of the flu is not the flu itself. While a brain illness is a physical illness, representing changes at the microscopic level in the brain, the spread of the brain illness among people can happen through various vectors of transmission, like trauma in the PTSD example above, or an economic crash, inefficient values, or genetics.

✳

We noted before that the physical impact of the information on the body is an emotion. The fact that emotions modulate synapses—the connections between neurons—is well established in psychiatry. It follows that the physical impact of the information on the body modulates synapses. In short, the information modulates synapses.

The synapses have memory. They know how to read the "traffic" of the information that goes through. As information goes through, the synapse increases in size. When no information goes through, the synapse decreases in size.

Not all information is under self-control. Unconscious information, in the form of unconscious thoughts, is not under self-control. However, like any information, unconscious information modulates synapses. There are synapses dedicated to just that: unconscious information. As synapses have a level of organization higher than randomness, synapses dedicated to the unconscious information take away the randomness from the unconscious.

We noted earlier that "I do not trust" gets to reign in the unconscious in PTSD. It cannot be simply turned off, being out of awareness. Instead, "I do not trust" is a thought that repeats itself unconsciously, within the well-known compulsive nature of the unconscious in general—discussed a long-time ago by Freud.

Where "I do not trust" repetitively goes through, the synapse grows in size. Then, new incoming information from the environment is lured through the growing synapse of "I do not trust", tending to associate the new incoming information with "I do not trust", beefing up the emotion of fear accompanying "I do not trust". By association with the previous traumatic information that went through the same synapse, the new incoming information *feels* familiar, the feeling taking the form of fear accompanying the underlying unconscious thought of "I do not trust". This is particularly effective when the new incoming information is not far from the original traumatic information.

Let's consider the example of a boy who had a traumatic experience with his verbally abusive mother and developed PTSD. Subsequently, non-abusive encounters with the mother tend to travel through the same synapse in the unconscious where the thought "I do not trust" is represented, as the boy does not have control over the association of the mother with "I do not trust" in the unconscious. Disconnected from conscious trust—due to the fracture in ego function in PTSD—the unconscious thought "I do not trust" is expressed through repetitive compulsion in a growing synapse in the unconscious. Moreover, the emotion of fear accompanying the repetitive unconscious thought "I do not trust" re-traumatizes the child, reinforcing the unconscious lack of trust.

Another woman facing the boy tends to travel in the unconscious of the boy through the same synapse where the unconscious thought "I do not trust" is represented, because the child associates the new woman with the mother, through similarities in perception. Thus, "I do not trust" expands to the new woman, with the accompanying emotion of fear too.

By cumulative effect over time, the process can lead to an expansion by generalization of "I do not trust" to women in general. This can happen especially when the mother continues to be verbally abusive, reinforcing the stereotype of "I do not trust" in the unconscious of the child. Also, other women being verbally abusive do not help either—for instance, chances are that the maternal grandmother is verbally abusive, like the mother.

Years later, meeting a potential girlfriend tends to travel in the unconscious through the same—now expanded—synapse, where the unconscious thought "I do not trust" is represented, by associating the potential girlfriend with women in general. Then an accompanying emotion of fear happens—this time, of the potential girlfriend. The lack of trust with the accompanying emotion of fear can result in rejecting the potential girlfriend, not because of anything about her, but because of everything about him. Consequently, the opportunity of the relationship with the potential girlfriend is lost.

This is when the unconscious thought "I do not trust" wins, as the past is re-experienced. The past is then re-memorized, further increasing the synapse where "I do not trust" is represented, expanded by now to women in general. Time travels then in reverse, turning future experiences into past experiences. The PTSD of the individual traumatized by the verbally abusive mother in childhood interferes with functioning years later.

※

The unconscious is a negotiator at the table of the present, because time is always now, in the present, in the unconscious. The past is not the past but is the present in the unconscious. Time is a matter of consciousness. Outside awareness of time, a person is always in the now. A consciously familiar situation *is* the previous situation in the unconscious. The conscious "this feels familiar" turns into "this is happening again" in the unconscious.

※

Neuroscience established that fear of trauma enhances the memory of trauma. But more: in PTSD, fear of trauma tends to follow the patient to the present. When recalled during psychiatric treatment, the memory of trauma feels uncomfortable in the present, not only back then. It makes the psychiatric treatment more difficult. The patient brings the fear of trauma to the patient-psychiatrist mind. The work in trauma therapy does not start from an emotional zero. Instead, it starts from fear: the patient *feels* afraid in the room with the psychiatrist.

As the fear goes beyond reason (patient is afraid despite knowing better), the "washout" of the fear needs to happen not at the rational adult level, but at the level of the emotional child within.

As the patient gains a sense of security in treatment, the inner child comes to surface. Even so, fear is still present. Because of it, the "child" looks for the source of fear in the present—where the "child" sees first the psychiatrist. Naturally all children attribute how they feel to what they see, so the "child" attributes the source of fear to the psychiatrist. The upside of this misattribution by projection is the lifting of the buried emotion of fear from deep inside to the conversation with the psychiatrist. Through the relationship with the psychiatrist, the patient can deal with the scared child within.

＊

Talk therapy is not so much about *what* is talked about. Instead, talk therapy is more about *how* the what is talked about: the underlying emotion matters more in therapy than the content of talk. In free associations during therapy, the associations are only apparently free—of the logic of the spoken words in the present; however, the associations are not completely free, being connected by an emotional thread across various points in time.

When witnessing an emotion on the part of a patient with PTSD, the psychiatrist has a decision to make: either follow the logic of the spoken words, or follow the emotional thread. The logic of the spoken words is not always there. For instance, let's say the traumatized patient feels the emotion of hate toward the psychiatrist. By following the logic of the spoken words, the psychiatrist runs into a logical barrier, because usually the logic of hating the psychiatrist is not there. By following the emotional thread instead, the psychiatrist opens the door to the unconscious of the patient, where the unconscious thought of "I do not trust" hides behind the emotion of hate—despite the conscious trust inherent to seeing a psychiatrist.

The patient with PTSD cannot directly deal with the unconscious thought of "I do not trust", because the patient is not aware of it. The patient is aware, however, of the emotion of hate, which operates like a tunnel from the unconscious to the conscious, touching on both (hate in raw form, rooted outside awareness, touches on the unconscious; and hate in projected form, on the psychiatrist, touches on the conscious).

Through a similar emotional tunnel, a new feeling finds its way in reverse, from the conscious to the unconscious: the emotional comfort in therapy. Note that the emotional comfort in therapy is more likely to travel from the conscious to the unconscious when the words spoken by the psychiatrist are centered within the detailed circumstances of the patient, with their twists and turns, instead of the surface of the general meaning of words (we already established that the ego is an individualized function negotiating between the conscious and the unconscious; thus, the words of the psychiatrist, which drive the emotional comfort in therapy, are more likely to resonate with the ego when they are individualized).

As hate and emotional comfort meet at the border between conscious and unconscious, a transaction of feelings occurs. When hate overcomes emotional comfort, the unconscious thought "I do not trust"—accompanying the hate—is elevated to the conscious, and subsequently the patient is looking at firing the psychiatrist. When, however, emotional comfort overcomes hate, the conscious thought "I trust"—accompanying the emotional comfort—is heading to the unconscious. Within the unconscious, the thought "I trust" gets now in the position to replace the thought "I do not trust," as part of the healing in PTSD.

Without first going through the emotional tunnel, a thought cannot pierce the boundary between the conscious and the unconscious, in either direction. The tunnel of feelings is necessary for a thought to travel from the unconscious to the conscious, as well as for a thought to travel from the conscious to the unconscious. The emotional tunnel is the way around the logical block stopping information from directly traveling between the conscious and the unconscious, in either direction.

✳

For a patient with PTSD to unlearn trauma, a replacement of the thought "I do not trust" by the thought "I trust" needs to happen in the unconscious. As the thought travels through a synapse, competition ensues between synapses: on one side is the synapse of where "I do not trust" travels, on the other side is the synapse of where "I trust" travels. The winner takes all, like for two parallel highways: when most traffic goes through Highway 2, it is Highway 2 gets money for reconstruction and expansion, while Highway 1 deteriorates and eventually becomes unusable; in the end, Highway 2 absorbs all traffic, overpowering Highway 1. Similarly, when "I do not trust" is the bigger thinking compulsion compared to "I trust", the synapse of "I do not trust" predominates over the synapse of "I trust".

But a belief in a bright future gives strength to "I trust", against "I do not trust." Note that the belief in the bright future does not have to be rational in order to give strength to "I trust". The belief can merely be assumed against reason, against the experience of what happened already. Through belief, hope when "I do not trust" becomes love when "I trust".

In a YouTube interview, the famous conductor Sergiu Celibidache of Munich Philarmonic in Germany talks about the ability to feel ahead of the feelings that would be generated by what happened already. Before even saying a word, the conductor exhibits a sense of confidence. As the conductor begins talking, his confidence unfolds toward the future, rather than the experience of what happened already. The enlightenment on his face together with his spoken words do the same thing as good music: squeeze into a corner the yesterday, putting tomorrow in a warm light.

Transcending from reactive feelings to feelings unrelated to reason is an opportunity to overcome the limitations of the physical existence. By not having to rely on the five senses, transcendence is left to the inner world of each person. When the inner world is sustained by belief, transcendence gains the strength of reality.

The ability to feel for the bright future believed in, as opposed to experiences that happened already, allows the soul to stop being a prisoner. Celibidache regards it as the moment of freeing up.

While competing thoughts are made of information and modulate synapses (similar to how traffic modulates two parallel highways), the accompanying emotions to the competing thoughts have a physical impact on the same body. What for thoughts is a competition between synapses becomes opposite simultaneous emotions in the same body.

✳

There are four primary sources of information: the physical environment, the mental environment, the genetic DNA software in the brain, and the conscious values software in the brain. The information from the four primary sources is processed by the brain, outputting a new resulting information. As we already know about information in general, the new resulting information has a physical impact on the body: an emotion, which can spread across both the conscious and the unconscious.

Because the brain processing of the information from the four primary sources can be less than perfect, sometimes, upon processing, old residual information lingers along. The old residual information has a physical impact on the body too—another emotion, which can spread across both the conscious and the unconscious.

The simultaneous presence of the two emotions (the emotion from the new resulting information processed by the brain from the four primary sources, and the emotion from the old residual information) gives birth to a conflict of emotions.

For example, a conflict of emotions takes place in borderline personality disorder developed in the context of being verbally abused in the childhood by the parent. Let's remark first that a victim of verbal abuse is not restricted to developing one illness only. For instance, a victim of verbal abuse, as noted previously, may develop PTSD; another victim of verbal abuse, as noted below, may develop borderline personality disorder. The causative agent being the same—verbal abuse—the disease is different—due to the vulnerability of the individual for one illness over another (e.g. genetic vulnerability).

Let's now delve in the conflict of emotions taking place in borderline personality disorder developed in the context of being verbally abused in the childhood by the parent.

Unconsciously, the child may emotionally pair by association love with verbal abuse, when the parent provides both at the same time. Remember the cartoons, when a newborn duck comes out of the egg and follows the first creature it encounters, which sometimes is a cat? The brain of the newborn duck is programmed—called imprinting— to attach emotionally to the first creature it sees. Later, the newborn duck will not follow the mother duck, but the cat. Through imprinting, the first being in sight *becomes* the mother. That is who the newborn duck trusts.

Likewise, when the first creature the child sees is the loving parent who is also verbally abusive, the child attaches emotionally to the love combined with verbal abuse. Trust takes place in the verbally abusive parent. When love without verbal abuse is absent, the child does not resonate with it, being unfamiliar with it. The presence together of love and verbal abuse creates an ambivalent emotion of "love" and unhappiness in the child, which is the emotion trusted by the child unconsciously later in life, by resonating with it.

Once grown, the victim projects the old ambivalent feeling of "love" and unhappiness on the romantic partner. This ensures the resonance of the feelings once had for the parent with the feelings for the romantic partner, now "loved" by the victim the way the victim knows how to love. In doing so, the future is molded unconsciously, to fit into the emotional experience of the past: "love" and simultaneous unhappiness to occur in the new relationship, inviting love-hate in. But when the romantic partner does not resonate with the love-hate proposed, the romantic partner is looking at ending the relationship.

Essentially the victim unconsciously recreates what the victim *feels* true love is—the same ambivalent emotion of "love" and unhappiness from the childhood. Consequently, the victim may be emotionally restricted to a romantic partner that resonates with a similar emotion once created by the parent in the victim: "love" and simultaneous unhappiness.

For the victim of abuse in childhood, pure love in adulthood, without unhappiness attached, would be a strange experience, it would not *feel* right—and therefore it is not trusted unconsciously, not being resonated with.

In the economics of emotions, it may be too "expensive" to modify emotions: a kind of love never felt before, unknown to the heart, and thus unappealing emotionally, does not *feel* worth the effort. For the victim to disconnect from verbally abusive "love" and look for love without verbal abuse attached, the victim must emotionally jump over the past, which is difficult.

For one, a barrier is that the emotional resonance takes place in the unconscious, where the past is ingrained. Plus, another barrier is that the verbally abusive parent had a good side too, making letting go of the good side of the parent inherently traumatic. Moreover, letting go of the "love" as known from the parent opens the door to the fear of unknown, which was never mitigated unconsciously through an experience of love without verbal abuse. When the fear of unknown arrives, a fight or flight reaction takes place, resisting change from what gives comfort—in our example, resisting change from "love" and simultaneous unhappiness.

If unable to emotionally jump over the past, the victim becomes the prisoner of a need for "love" and simultaneous unhappiness, in order to resonate emotionally. Being a prisoner is the disorder: at the border between "love" and simultaneous unhappiness, the borderline personality disorder is born.

Looking at what unconscious thoughts underlie "love" and simultaneous unhappiness, the core of borderline personality disorder is the coexistence of the unconscious thoughts "I trust" and "I do not trust", in a tension blocking either from taking over the other (in healing, "I trust" takes over the "I do not trust"; in PTSD, "I do not trust" takes over "I trust").

Being without cure, nothing short of a miracle in borderline personality disorder can dissipate the unresolved tension blocking against each other the unconscious thoughts of "I trust" and "I do not trust".

FAITH

As science, psychiatry looks for verifiable, measurable results. On the other hand, stepping outside the results of psychiatry, faith counteracts the decaying nature of the visible world. Faith is a plunge into the assumption that the future is good, independent of what is going on now. Faith does not see the present as a predictor of the future, like psychiatry does, but assumes a good future despite the evidence. Faith has the ambition to contradict the verifiable, measurable results of psychiatry.

How can then a modern psychiatrist relate to faith, when the modern psychiatrist is a disciple of science? In order for the psychiatrist to plunge away from science, a paradigm shift is required. While in general, the psychiatrist is called to assist people with broken brains, the people with broken brains beyond repair fall outside the reach of science. To find a new lens for patients with broken brains beyond repair, the psychiatrist is called to look farther than science, to look where hope shines *beyond* the here and now. The psychiatrist must go from being "a warrior against adversities" to being "a painter of the beauty ahead".

For the warrior to coexist with the painter is no easy task for the psychiatrist. On one side, the psychiatrist must wield a sword to cut through adversities: the pen, to write medical solutions with. On the other side, the psychiatrist must use a paintbrush to envision a bright future: the same pen, to anticipate where life is going. Then, the loss of the patient's control over the decaying world turns into a grasp of control over an imagined future.

Faith has long unleashed the colors of a bright future against a grim reality. People who have faith owe to those who live in the shadow of brain illness to open the window of imagination together.

✷

On the question of whether there is life after death, not only the imagination gives an answer, but mathematics too. Here is how.

While there is no proof of life after death, a mathematical truth beyond doubt is that any possibility of the infinite is larger than the certitude of zero. Therefore, a possibility of the infinite by a leap of faith that there is life after death is larger than the certitude of zero. Consequently, believing there is life after death is mathematically worth.

✷

This book refers to faith from time to time. I'd like to take a moment to distinguish faith from fanaticism. Anti-faith people should thus be precluded from simply grouping together faith and fanaticism for the purpose of counter-arguing the points about faith in this book.

People kill people in the name of religious belief. These killings cannot be reconciled with the modern value of not killing others. Apparently, the words "but it is my religion" are used as a shield, hiding fanaticism on behalf of "religious" freedom, not faith. While a leap of faith that brings healing is good, a leap of "faith" that brings revenge and destruction transgresses into fanaticism, the realm of evil. The reference to faith in this book is limited to a faith that is good, in the sense of being consistent with foundational public values, such as not killing others.

To allow the mere words "but it is my religion" to cover anything can be dangerous. Assessing what's behind the words requires an interaction with the "religion", for instance by observation and dialogue. The "religious" belief must pass the muster of a minimal decency, in order to be consistent with foundational public values, such as not killing others.

Fanaticism can falsely resemble faith in the rigidity of being undisputable to the believer, but fanaticism differs from faith in the substantive values it promotes. When a "religion" calls, for instance, for killing others, the public needs to step in and draw the line. Hate masked by "religion" is not worth being protected by the secular liberty of religious freedom.

✳

A patient once asked a psychiatrist: "Where is Heaven?" Tempted to reply, the psychiatrist stopped short of answering—because where Heaven is begins by imagining what Heaven is like. To the question "Where is Heaven?" the psychiatrist turned the table to the patient and said: "How do you imagine Heaven to be?"

The inner journey to the roots of imagination is of emotional nature. The imaginary Heaven is personalized for each individual, populated with who is meaningful and dear to that person. For believing that imagination has no underlying roots is a mistake. Every serious psychiatrist knows that the imagination is not a free-floating alternative to reality. Instead, imagination is rooted in reality.

When brain illness fuels disbelief in the reality experienced so far, suspending the disbelief is a therapeutic form of moving forward. The psychiatrist can help the patient suspend the disbelief, through the lens of the therapeutic relationship. With disbelief suspended, the future can break free from the past.

ART

The concept of suspended disbelief comes up in relation to art, for instance the art of motion pictures. Enjoying a good movie uses suspended disbelief to disregard that the movie is fiction. For instance, when a character shoots another character that dies from the gunshot wound, no true shooting actually takes place.

While the reason tells the spectator that the movie is fictional, the emotions felt when watching the movie are real. The spectator truly feels for the characters on the screen. *Real* feelings happen when disbelief is suspended.

Feelings within the spectator cannot be faked: a person either has the feelings or not. On the other side, to *know* that a show is fictional despite *feeling* real takes the skill of reasoning. When suspending disbelief, the spectator turns off reasoning in order to feel *real* emotions unencumbered by reason. This draws in an audience of millions.

Under the magic of the movies, the audience lives inside an unfolding story. The core need to belong of the audience is momentarily satisfied. But more: a good story pushes the boundaries of reality.

Let's take the example of *Star Wars*. The movie delivered a narrative of traveling to a galaxy far, far away. The story turns real when people are buying memorabilia from it, playing video games on the theme of the movie, or traveling to California to meet the movie characters from *Star Wars* at Disneyland.

By the way, Disneyland itself is *real*, despite being based on a fake mouse, because the emotions stirred by the little imaginary creature have been *real* for years. And there goes the conclusion: suspended disbelief expands reality. Now people belong not only to a fictional narrative in front of a screen, but to a real narrative by immersion in a magic world come true.

Likewise, a session with the psychiatrist invites the patient to develop real feelings toward the psychiatrist, and, through them, to be inspired by the psychiatrist. Besides projecting old feelings from previous relations on to the psychiatrist, also known as transference, brand new feelings for the psychiatrist develop, making the relationship with the psychiatrist the vehicle of change for the patient. Then, the right question from the psychiatrist can stir enough interest that the patient searches for an answer outside of what the is already known to the patient. This paves the way for the patient to expand reality. The expanded reality pushes the patient one step closer to taking back control over brain illness.

✸

The ambivalence of a question posed by a good movie allows the seed of art to grow in the souls of the people watching the movie. When spectators compare on-screen lives with their own lives, discussion ensues.

In general in art, a good artist does not deliver a full answer, but leaves a question hanging. When people try to resolve the riveting question raised by the artist without a set answer, creation happens. By stirring up a conversation, the artist stimulates an interest in the topics surrounding the artistic seed.

Some of the best works of art reveal ambivalence, leaving the audience to resolve a question. For instance: Is Mona Lisa smiling or frowning? In fact, the artistic value lies in the unanswered nature of the question. It gets people to talk, come up with answers, project feelings. The self-exploring of the audience's inner world is brought to light by the ambivalence of the artistic question posed.

From time to time, the psychiatrist too poses intentionally an ambivalent question. When reasoning alone falls short in determining the answer to the ambivalent question, the patient is invited to use feelings in order to look for an answer. As nobody *knows* the right answer, the patient is left to merely *feel* the answer. The ambivalent question becomes a tool to explore feelings, some of which are rooted below the surface of consciousness. Bringing feelings under the light of consciousness opens the door to self-exploration beyond what the patient already knows about oneself.

When much of what the patient answers in the psychiatric session is driven by unconscious feelings, outside self-control, the patient has an illusion of control over how the session goes. The psychiatrist walks a fine line to create an illusion of control for the patient over feelings, while at the same time using feelings outside the patient's self-control to get the patient to travel to the unconscious.

✳

In Michelangelo's famous painting "The Creation of Adam", God offers a hand to Adam in the Garden of Eden. In turn, Adam begins by looking at God. The connection is a symbol of the mind between God and Adam.

Once Adam later turns away from God, Adam implodes within his earthly nature. Adam is the victim of an illusion of choice: of being able to let go of God, which takes away the power of the mind formed with God.

The illusion of choice is prevalent: people think they have control over their lives. However, the variables are too many to control. Every day with a peaceful outlook is a blessing from God. Being thankful for it is reasonable.

✳

When nobody is left around to form a mind with, society pays a psychiatrist to pop up. And yet, God calls the brain of the human to form a mind with God foremost, not with the psychiatrist. Thus, the newly popped-up psychiatrist is in a rather odd position, between God and the human.

In the desert of brain illness, the patient looks for an answer. When the answer does not come from within, the patient searches for the answer by way of the mind. This raises a moral question: will the psychiatrist inadvertently stay in the way of the patient forming a mind with God?

That the psychiatrist can choose to remain neutral is an illusion. The psychiatrist steals the attention of the human and, therefore, carries the responsibility to not compete with God.

✳

Art museums show plenty of marble rocks. At first, every rock is a cube, a block, a stone with no meaning. Then, someone shapes the rock into the most exquisite statue of a human head, with eyes, nose, ears, and even hair over the head. It stares with meaning, showing affect, and appears to be thinking. It's neat, smooth, and almost alive. The brain that sees the statue connects with the brain of the sculptor, through the rock—now a fully-fledged piece of art.

Likewise, the psychiatrist connects with a sculptor stepping into the office every hour, that works on a rock shaped by ideas, dreams, wishes, losses, and shadows of memories. Every rock has the potential to become a piece of art. Patients reveal life stories in front of the psychiatrist from behind words, through stages of emotional waves.

For most people that see a psychiatrist, something is not working out in their sculpture. For instance, a part is missing, or is too big, or is too small.

The job of the psychiatrist is to compare the sculpture created by the patient with a more beautiful version of it. As the patient is blindfolded by illness, and knowing that beauty is in the eye of the beholder, who is the beholder? By necessity, the psychiatrist becomes a substitute beholder for the patient.

The psychiatrist has a great responsibility, being called upon to see the richness of the creation beyond what has been created already, the potential of a future not manifested so far.

✳

But is the psychiatrist not merely a tool, to take the patient from brain illness to brain health?

Amongst psychiatrists was once said that the better the psychiatrist, the more likely the psychiatrist is to get along with all sorts of patients. Engaging difficult patients in treatment requires a capacity to see through what happened already, to a better future. A good psychiatrist can help patients compress the shortcomings of life experiences into a better reality of tomorrow.

It's like repairing clocks. When finding a tiny piece of metal on the floor, most people probably don't know what to do with it. But someone that repairs clocks may envision a clock where the tiny piece of metal belongs.

For a better reality of tomorrow, the psychiatrist needs a vision on the world. By admiring the beauty of where the lost belong, the psychiatrist can use the vision to articulate back the functioning of the patient, in synchrony with other people.

As some patients are ill enough to require help from the psychiatrist for an indefinitely long time, the psychiatrist has the opportunity to walk the walk: to make a difference not merely in how these patients perceive what goes on, but in sustaining their function in the surrounding environment. In that, the psychiatrist goes beyond the role of a tool, becoming part of the sculpture.

TURNAROUND

When these words go from pen to paper, the temperature on the outside is zero degrees Fahrenheit. Truly cold. The pain of the winter hits with force. One can feel it in every bone, every fiber, every corner of the soul.

The pain of the dysfunction for a patient can feel like the zero degrees Fahrenheit. It feels *for real*, in the objectivity of the subjective experience of the patient. Not knowing where to turn, the patient trembles and shivers. Traveling to the future failed. Going back to the past, through the brain illness, is what's left. Like a boomerang, the patient attempts to travel forward, only to go in reverse.

And yet, more discoveries take place in neuroscience every year. Technology is now over-the-top advanced. When restoring functioning, the psychiatrist facilitates the patient trying to break the hold of the past on the future.

✳

In the mental health wing of a large jail, inmates are tense, psychologically tortured by the loss of freedom, staring through the ceiling to a sky hidden from sight. A frail, barely moving, elderly nun walks the never-ending hallways. She goes from cell to cell, engaging inmates in conversation. Without medication, is that treatment?

Beyond the walls of the material universe is the power of words. Through the words, the material universe becomes integrated into relationships. Moreover, beauty lies in the ability to create together, an adventure in which people conquer the material world for the sake of relationships.

For instance, a teddy bear from a grandfather is an inanimate object, without biological life. But the teddy bear carries the spirit of the relationship with the grandfather. This turns the teddy bear into a treasure, alive psychologically, by reflecting the life of the relationship with the grandfather.

Brain illness needs to be talked to. For a long time before medications, a psychiatrist had only words to work with. Pills came later. Pills disentangle the soul from the swamp of biology, letting the spirit soar into the sky of connections with other souls. But on the question of whom to connect with, the answer is not coming from the pills.

✳

Due to professional boundaries in place for the rightful protection of the patient, the connection between the patient and the psychiatrist is limited. For instance, the connection is limited to an exchange of money for psychiatric services, like prescribing medications. For a deeper connection, the psychiatrist is not in a good position. Instead, the patient is left with a need to form a mind with a third party.

Who can be the third party? In a good position to be the third party is, for instance, a churchgoer—if the churchgoer would only step up to the plate, knowing that the psychiatrist is around too. In doing so, the churchgoer can stop trying to save oneself, and become instead an opportunity for the lost soul to connect with.

When bringing light to the dark, warmth to the cold, and song to the silence, the churchgoer gives a narrative to the lost soul to belong to. Through acts of kindness and mercy, the faithful can help fulfill God's hope that people with brain illness enjoy life in His spirit. Note that the connection between the churchgoer and the patient is not limited to words. Acts of good deeds by the churchgoer can open doors of the patient's soul that are closed to words due to brain illness.

✸

For churchgoers, to give a chance to a patient with brain illness may be hard. It may be as hard as, let's say, not eating meat. Both go against the instinct of survival.

Churchgoers attempt to find salvation by forming a mind with God. That brings them a sense of power. But what do they do with the power, when it comes to powerless beings? As for the relation between the churchgoers and animals, killing animals like in the Stone Age is not really the way to show love for the creation to God. For what does God see in the eyes of the animal about to be killed by the churchgoers?

Today, plenty of sources of food are available that do not involve killing animals. With so many vegetarian choices, eating animals is a mere convenience of the powerful over the powerless. Oddly, the powerful can be the same people who ask God for mercy at their weekly religious service. Just because God gave them power over animals does not mean the animals are here to be killed. The mercy asked from God is the same mercy that can be shown to God in the form of letting the animals be. The Nobel Prize winner by the name of Isaac Singer wrote about this poignantly.

Hopefully, engaging churchgoers in a conversation on this topic can increase their sensitivity to developing empathy for the animals, to the point of sparing the lives of the animals.

As for the relation between the churchgoers and the people with brain illness, churchgoers do not literally kill people with brain illness. Instead, they reject people with brain illness often—a killing of the mind. Moreover, people with brain illness, ashamed of their failures, are sometimes shy of the accomplishments of churchgoers, and afraid to be judged in church. Consequently, people with brain illness might stop short of entering the door of the church.

How could this change?

A church more educated about brain illness refrains from judgment and extends a helping hand. Being a churchgoer is not a race to a finish line, but a search for a fresh start. "Where do I end up?" is not the question. "Where do I begin?" it is.

＊

As noted already, dysfunction usually comes by definition with brain illness. Consequently, people with brain illness tend to fall short of the expectations of churchgoers. Trust offered by the churchgoers, when stolen by people with brain illness, gives no apparent return.

Without return, the churchgoers are inclined condemn people with brain illness to isolation by rejection, crucifying them on the cross of loneliness. No mind is left on the cross, but merely an ill brain, rejected.

Under the rejection of the churchgoers, people with brain illness are in the predicament symbolized by the two thieves on the cross on the day of the crucifixion of Jesus. One thief went on the path of distrust, despair and negativity, staying mindless. The other thief pressed forward with the trust that not everything is lost, despite the evidence, and searched for a mind—finding the only mind left, that with the innocent on the cross.

Likewise, every rejected person with brain illness has the Seven Words of Jesus on the cross to relate to.

The First Word: "Father, forgive them, for they do not know what they do". In praying to forgive the humanity for rejection, Jesus reminds the Father how humans do not connect their action ("what they do") with what they know ("they do not know"), instead being driven in action by a factor outside of their knowledge. In praying to forgive the churchgoers for rejection, the person with brain illness offers to God the trust stolen from churchgoers, hoping to heal the broken relation with the churchgoers, which are driven to rejection from outside of their grasp.

The Second Word: "Truly, I say to you, today you will be with me in Paradise". Here, the power of the mind overcomes death, after the penitent thief offered trust in Jesus. Death has no place in Paradise. When death is not the end of the road, it follows that brain illness is not the end of the road. Note the use of the word "truly" by Jesus, emphasizing how the truth is not dependent on the evidence.

The Third Word: "Jesus said to his mother: 'Woman, this is your son'. Then he said to the disciple: 'This is your mother'". Family is not restricted to the biological family. Those who form a mind with God are family, despite the appearance to the contrary.

The Fourth Word: "My God, my God, why have you forsaken me?" Here, the mind between Jesus and the Father breaks down. Jesus, while being only human, projects on the Father the distrust inherent to human nature: "Why have *you* forsaken me?", instead of "How do *I* trust you?". The trust is gone, and Jesus, only a human then, perceives the silence of the Father as a giving up by the Father. But it is not for the Father to follow the human in hell. It is for the human to find the way out of hell, back to God. Christ shows the way, next.

The Fifth Word: "I thirst". The thirst is for more than water. It is for connecting with lost people—only to be met by the Romans with sour wine. The penitent thief, however, turns to God, offering the trust he stole from society to God instead. By asking to be remembered in Christ's kingdom, the penitent thief offers the water that Jesus was most thirsty for on the cross.

The Sixth Word: "It is finished". Arguably, Jesus is referring not so much to the sacrifice of the body, but to that of the mind between Jesus and the Father, not even such a sacrifice being enough to earn the trust of the impenitent thief. Condemned to the cross of isolation, the patient has a choice: faith versus evidence, symbolized by the two thieves on the cross. In the absence of the evidence of God's power, released to the human by God, trust is built on faith.

The Seventh Word: "Father, into your hands I commend my spirit". This is the moment when Jesus, while still only a human, renews the mind between Jesus and the Father, broken down not long ago. Two minds come to life on the cross: the mind between the penitent thief and Christ, and, separately, the mind between Christ and the Father. By bringing the trust from the penitent thief to the Father, Christ becomes the bridge between people who stole God's trust and God. Christ becomes the way for the lost people back to God, replacing reason with faith. Each person with brain illness has a choice to make, like the two thieves on the cross once did: apply reason to evidence, or release the control of reason to faith despite the evidence.

People with brain illness, on the cross of isolation, unable to control rejection by churchgoers, may release the control of reason to faith despite the evidence. Then people with brain illness and churchgoers can meet in faith before meeting in trust, as faith can guide the churchgoers to the innocence of caring for a lost soul.

While waiting on the cross of isolation for the churchgoers to show up, distrust and faith are in tension within a person with brain illness. The distrust is deep, in the unconscious, where one cannot reach by direct conversation with the self. A conversation outside the self is needed, to chase the distrust away. Like Christ once did, the person with brain illness has God to turn to. Another name for it is *praying*.

✳

Centered on the mind, the science of the brain reaches a limit. While the brain is subject to science, the mind goes beyond it.

If Pilate would have sent Jesus for evaluation of competency to stand trial by a psychiatrist, chances are quite high that the psychiatrist would have diagnosed Jesus with major depression with psychotic features, in light of the readiness of Christ to be sacrificed. This is the predicament psychiatry is in today: the optic of the lens changes perspectives. With a Christian lens on, what may have previously seemed delusional becomes now normal.

From a Christian perspective, carrying the cross is a healing process, despite suffering. Simple acts, like working long hours to pay a kid's college tuition, or providing for an ill relative, or giving time to charity, are altruistic behaviors in which the brain suffers but the mind gains. The transfer of power from the brain to the mind gives power to the mind to reach values that are counterintuitive to the brain. While at first the brain is left powerless on behalf of the mind, the values only the mind can reach return power, through the mind, to the brain.

In other words, the brain is not alone anymore.

For what is a value after all? A value is the stored expression of a preexistent mind, to which a new mind can relate. The relating of the new mind to a preexistent mind pulls the new mind in directions neither of its two brains do—the new mind gaining a life of its own, independent of the two brains. The gain is enriching connection, giving the new mind the characteristics of an independent person, not dependent on the two brains anymore.

✻

Christianity describes "one God in three Divine Persons": Father, Son, and the Holy Spirit. Interestingly, God is not the Father, nor the Son, nor the Holy Spirit, but is the three Persons together. While one can relate easier to the concept of "the Father" and "the Son", standing for the more classical understanding of what a person is, the concept harder to relate to is "the Holy Spirit", which stands for the mind between the Father and the Son. The mind gains a life of its own, independent of the Father and the Son, when the mind relates to other minds, pulling it in directions neither the Father, nor the Son do. The gain is enriching connection, giving the mind the characteristics of an independent person. This is illustrated by the release of the Holy Spirit into the world after Jesus left Earth to ascend to Heaven.

On a simpler level, when two spouses form a family, asking God in the act of wedding to bless the new family with sustaining life, the mind between the two spouses is about to get a life of its own. Not long after, the family reaches a unique expression of the mind between the two spouses: the child. The mind takes the form of a body, with an independent life of its own.

Note that the Bible says, through the Apostle John, "In the beginning was the Word". In the book of Genesis, God first said the Word—or spoke His mind: "Let there be ..." Then, God created the human. It appears that the mind preceded the brain—with the brain being the product of the mind, not the other way around.

＊

As an expression of the mind of doctors, psychiatry is called upon to not reduce itself to serving brains, but to go beyond, to the realm of minds. By virtue of the mental connection between patients and God, psychiatry is invited to converse with God for the sake of the patients.

Stepping outside the box is key to go from the breakdown of psychiatry to the sustainability of psychiatry.

＊

I've talked with people for whom the mere mention of Christianity was perceived as a threat to their fabric, security, and sense of self. They look at me as if I'm delusional and that I should know better than to speak in public about my delusion. My belief in Jesus indeed makes me a target for rejection by many. Did I mention ridicule, too?

And yet, I continue to believe, and I hope that we can get along in the future. For nonbelievers, to appreciate what Jesus did is within the range of reason. While many people can coexist by reasoning, some use their hearts to believe.

A delusion, being religious or not, simply indicates the formation of a strong neuronal node that functions with a certitude of truth for the individual. Not all delusions are bad. Plus, the delusion can be in the eye of the beholder. The reality may belong to the delusional person, while imagination is what's left to the realistic individual. After all, what is the love to conquer reality, but a delusion of power?

The shared reality of a patient that says "God, help me" with the churchgoer that says "God, let me do an act of grace in your name" can overcome the delusional scientist that says "I am studying psychiatry to figure out how the brain works". Because it is not the brain that does the work, but the mind.

www.ingramcontent.com/pod-product-compliance
Lightning Source LLC
Chambersburg PA
CBHW022108280326
41933CB00007B/305